PERSONALIZATION

PRINTING INDUSTRY CENTER SERIES

The New Medium of Print: Material Communication in the Internet Age
  by Frank Cost, 2005

Data-Driven Print: Strategy and Implementation
  by Patricia Sorce and Michael Pletka, 2006

Print Media Distribution: A Look at Infrastructure, Systems, and Trends
  by Twyla Cummings with Bernice LeMaire, 2008

PERSONALIZATION

DATA-DRIVEN PRINT AND INTERNET COMMUNICATIONS

PATRICIA SORCE, PH.D.

RIT CARY GRAPHIC ARTS PRESS
ROCHESTER, NY

Personalization: Data-Driven Print and Internet Communications
Patricia Sorce, Ph.D.
Printing Industry Center Series: 4

Copyright © 2009 Rochester Institute of Technology,
Cary Graphic Arts Press, and Patricia Sorce. All rights reserved.

No part of this book may be reproduced in any form or by any mechanical or electronic means without written permission of the publisher and/or individual contributors, except in the case of brief quotations embodied in critical articles and reviews.

Published and distributed by
RIT Cary Graphic Arts Press
90 Lomb Memorial Drive
Rochester, New York 14623-5604
http://carypress.rit.edu/

Printed in the United States by Thomson-Shore, Inc.
ISBN 978-1-933360-40-9

Designed by Marnie Soom / RIT Cary Graphic Arts Press
Typeface: Whitney by Tobias Frere-Jones / Hoefler & Frere-Jones
Excerpts reprinted by permission.
All trademarks are property of their respective owners.

    Library of Congress Cataloging-in-Publication Data

Sorce, Patricia A., 1950–
    Personalization : data-driven print and Internet communications / by Patricia Sorce.
       p. cm. — (Printing industry center series ; 4)
    Includes bibliographical references and index.
    ISBN 978-1-933360-40-9 (alk. paper)
    1. Printing industry—Marketing. 2. Printing industry—Customer services. 3. Printing industry—Management. I. Title.
    HF5439.P7S67 2009
    686.2068'8—dc22

                                                       2009029775

CONTENTS

List of Figures vii
List of Tables viii
About the Author xi
Foreword xiii

1 The Power of Personalization. 1
    Personalization Catches The User's Eye 3
    *Case Study: Gannett Co. Automates Multi-Channel
    Direct Marketing Program for Newspaper Renewals* 9
2 Personalized Marketing Communications in the
    Integrated Media World 17
3 The Current State of Personalized Marketing Communications . . . 37
    Hello, You 40
    Personalization Pitfalls of Internet Search 48
4 Personalization Strategies for Customer Development 53
    Ace Hardware Insert Test Passes with Flying Colors 58
    Appendix 4A: Overview of Relationship Marketing 69
    Appendix 4B: Web-Enabled Print Architectures 71
5 Database Technologies for Personalization 85
    WideWaters Gaming Case Study 90
    Appendix 5A: Database Fundamentals 99
    Appendix 5B: Overview of Four Variable Data Printing Products. . . . 108
6 Corporate Communications: In-Plant Print Shops and
    Transpromotional Documents. 119
    Evolution of an In-Plant: A Case Study of Printing Services at RIT . . 121
    Case Study: Ameriprise Financial Statement Redesign 139
    Case Study: First Data: The Opportunities are Limitless. 143
7 Transforming Printers and Publishers
    into Digital Service Providers 151
    Case Study: Thomson Corporation 154
    Case Study: A Newspaper Transformed.. 156

	Case Study: Global Printing, Inc.	163
	Case Study: Standard Register/Dealer Office XPress (DOX)	170
8	Measuring Success: Closing the Feedback Loop	177
9	The Intelligent Use of Personalization	197
	Case Study: Custom Publishing	204

References. 207
About the Printing Industry Center at RIT 221
Index 223

LIST OF FIGURES

Figure 2.1 Sample marketing plan for new shampoo line extension . . . 20
Figure 3.1 Annual amount of personalization used, 2003. 38
Figure 3.2 Customized magazine cover example from Spring 2008 RIT *Reporter* 39
Figure 4.1 A conceptual framework for CRM strategy 54
Figure 4.2 Coldwell Banker customized direct mail postcard 60
Figure 4B.1 Graphical representation of system analysis instrument . . . 78
Figure 5.1 A personalized campaign workflow 88
Figure 5.2 The personalized booklet campaign planning page 93
Figure 5.3 First page of personalized booklet 94
Figure 5.4 Personalized interior pages of booklet. 95
Figure 5.5 PURL survey sample 95
Figure 5.6 Campaign dashboard. 96
Figure 5.7 Automatic survey response notification form 97
Figure 5.8 Follow-up mailing to non-respondents 97
Figure 5A.1 The basic structure of a relational database 100
Figure 5A.2 A single record in a data file. 101
Figure 5A.3 Examples of data files in a relational database. 102
Figure 5A.4 Example flowchart of a nested IF statement 104
Figure 6.1 Annual revenues for RIT Central HUB print shop, 1988–2004 123
Figure 6.2 Annual impressions for RIT Central HUB print shop, 1988–2004 123
Figure 6.3 Revenues from RIT copy centers (charge-backs for both Central and Crossroads HUBs, and cash transactions from Crossroads HUB) 125
Figure 6.4 Variable data postcard developed by RIT students 131
Figure 6.5 An example of the "Preferred Application" variable data mailer 132
Figure 7.1 Transitional business model for the Democrat and Chronicle. 158

Figure 7.2	The *Democrat and Chronicle*'s transition from product to audience	159
Figure 7.3	The *Democrat and Chronicle*'s in-depth focus on audience	161
Figure 7.4	DOX variable data mailing piece	173
Figure 8.1	Four-year LTV calculation of static versus personalized catalog campaigns	194

LIST OF TABLES

Table 2.1	Primary services provided by advertising agencies	21
Table 2.2	U.S. advertising expenditures by media type	22
Table 2.3	Factors driving the media choices in a campaign	23
Table 2.4	"Below the line" media expenditures, 2002, 2007	24
Table 2.5	Direct response advertising expenditures in 2006	25
Table 2.6	Top personalization techniques	26
Table 2.7	Average response rates by direct medium	31
Table 2.8	Media that provides useful information on bargains	32
Table 2.9	Decision makers' top B2B digital media, 2007	33
Table 2.10	Decision makers' top B2B traditional media, 2007	33
Table 3.1	Amount of personalization used by marketing executives, 2003 vs. 2008	41
Table 3.2	Levels of complexity used in personalization campaigns	42
Table 3.3	Awareness of personalization print technologies	44
Table 3.4	Database functions and level of personalization used	46
Table 3.5	Correlations with level of complexity	47
Table 4.1	Hierarchy of dialogues	63
Table 4.2	Levels of personalization and the typical marketing objectives	67
Table 4B.1	Tabulated results from site analysis	80
Table 5.1	Relational information processes	86
Table 5.2	WideWaters Players' Club database	90
Table 5.3	WideWaters visit activity database	90
Table 5.4	WideWaters database provided to SourcePrint	92
Table 5B.1	List of variable data software programs	108
Table 5B.2	VDP application matrix: Software features and definitions	109

Table	Description	Page
Table 5B.3	Pageflex analysis	111
Table 5B.4	Darwin VI authoring tool analysis	113
Table 5B.5	PrintShop Mail analysis	115
Table 5B.6	XMPie uDirect Classic analysis	117
Table 7.1	Ancillary services offered by printers	153
Table 7.2	New styles for newspaper leaders	157
Table 8.1	Media metrics and related syndicated services	182
Table 8.2	Typical advertising impact metrics and related syndicated services	183
Table 8.3	Comparison of two media choices for a hypothetical advertising campaign	186
Table 8.4	RPM and CPR for an outcome of 100 dinner patrons	187
Table 8.5	Hypothetical results from static/mail merge catalog campaign	188
Table 8.6	LTV calculation for hypothetical static/mail merge catalog campaign	189
Table 8.7	Four-year LTV calculation for hypothetical static/mail merge catalog campaign	190
Table 8.8	Results from hypothetical static and personalized catalog campaigns	191
Table 8.9	One-year LTV calculation of hypothetical static and personalized catalog campaigns	192
Table 8.10	Four-year LTV calculation for hypothetical personalized catalog campaign	193
Table 9.1	Effective personalized communications tactics	199
Table 9.2	Three ways to build customer equity	201

ABOUT THE AUTHOR

Patricia Sorce is currently co-director of the Printing Industry Center at RIT and administrative chair of the RIT School of Print Media. She teaches in the areas of marketing research, buyer behavior, and database marketing. Her current research is focused on the topics of relationship marketing and the factors that impact the demand for data-driven print.

She earned a Ph.D. in cognitive psychology from the University of Massachusetts. She has published in refereed marketing, management, and psychology journals. These publications have spanned a wide range of topics including basic research in psychology, marketing segmentation analysis, and most recently, Internet buying behavior and relationship marketing. Before her appointment as co-director of the Printing Industry Center, she served as associate dean of the RIT College of Business from 1996 through 2001.

FOREWORD

IN THE THREE YEARS SINCE THE PUBLICATION OF *Data-Driven Print*, the media world has continued to change. A tangible example of this change occurred in late 2006 when Time magazine announced its Person of the Year—YOU! This decision on the part of Time's editors celebrated the growth of user-generated content and the enthusiasm of younger media users in having information "my way." But electronic user-generated media options such as social networking, downloading music to cell phones, file sharing, and text messaging have left mainstream media businesses such as news organizations, publishers, and record companies wondering how to make money.

To some, the mainstream media seems archaic. Take newspapers—they print news that comes off the wire services like the Associated Press and Reuters. These same wire services feed information to Internet news sites that make it immediately available to the world. If several years ago the morning newspaper seemed outdated compared to the previous-night's television news coverage, how much more outdated is it today, compared to Yahoo!® News? And as for television, TiVo and other digital video recording devices have given users control of their time and the ability to skip over or fast-forward through commercials. This has thrown the advertising industry for a loop: the mainstream media now reaches fewer viewers, and young media users avoid it altogether. The double-digit growth of search advertising, pay-per-click and keyword advertising over the past five years reveals where advertisers see the future in reaching prospective customers. Advertisers are taking advantage of Web 2.0 by creating their own blogging sites (such as Kodak.com), where they invite users to share experiences with the firm's products, success stories, and problem-solving tips. Some of these Web sites offer RSS feeds to committed consumers so that they get the latest news on products, services and other commercial content that advertisers paid for in the past. The ability to communicate 1-to-1 with interested individuals has changed the way businesses advertise, enabling them to concentrate more on customer retention and relationship building. This book is set within this context. Here the reader will find information on media advertising expenditures and the use of personalization as an advertising tactic. We review the latest thinking on relationship marketing and customer strategy development, and present information about electronic forms of personalization through email and Internet search. We also look at corporate communication beyond advertising, including how database and digital technologies have impacted in-plant printing and transactional communications. Finally, we discuss how traditionally print-centric firms are changing, and present several case studies to discover their best practices.

I would like to thank several people for contributing content to this book. My former students Adam Dewitz and Tracy Destino agreed to contribute their work, and current RIT student Vince Garguilo was helpful in identifying the latest information on Internet advertising metrics, summarized in Chapter 8. I also want to thank Barbara Pellow of InfoTrends who, along with fellow InfoTrends analysts Adam Peck and Matt Swain, provided updated research for this edition. Two recent case studies rely on the work of the Print on Demand Initiative (PODi). I want to thank Jon Budington, CEO of Global Printing, Inc., and John Meyer, director of RIT's "HUB" print/postal services, for sharing their stories with me. My appreciation goes also to my colleague at RIT, Twyla Cummings, Ph.D., for her excellent case study on change at Rochester's Democrat and Chronicle newspaper. Thanks go to Leslie Sprigg from Standard Register for an update of the DOX case study featured in Data-Driven Print. I would also like to recognize my former co-author, Michael Pletka of Xerox, for allowing me to include his work on database fundamentals. And I'm grateful to Pat McGrew of Kodak for thoroughly reading an early version of this manuscript.

I would not have been able to compile this book the help of the dean of RIT's College of Imaging Arts and Science, Joan Stone. Her continued commitment to the Printing Industry Center at RIT has provided a strong base for its research activities. Thanks also to my friend and co-director Frank Cost who provides the vision and energy to help the Center thrive. I also want to recognize Ashley Walker, the communications coordinator at the Printing Industry Center, who is chiefly responsible for the document management tasks that support this book. Her comments are always timely, accurate and insightful. I'm grateful to Liz Dopp, editor of the Center's online Print-in-the

Mix (a clearinghouse of research on print media effectiveness), for help with fact-checking and references, and to our publishing team at the Cary Graphic Arts Press, led by David Pankow. It's wonderful to work with such great people and dedicated professionals. Special thanks go to Molly Cort, Marnie Soom and Patti Cost for their expert help in editing, design and production.

I would also like to thank the Sloan Foundation and our Industry Partners for their support of the Printing Industry Center at RIT. These include Frank Mayadas and Gail Pesyna of the Sloan Foundation and the following Industry Partners who have been instrumental in our success: Adobe, Avery Dennison, Hewlett-Packard, Kodak, NewPage, NPES, the Scripps Howard Foundation, Standard Register, the U.S. Government Printing Office, Xerox, and the Democrat and Chronicle. They not only provide the financial lifeblood of the Center but also give us access to their firms for data collection and insightful commentary.

> Patricia Sorce, Ph.D.
> Administrative Chair, School of Print Media
> Co-Director, Printing Industry Center
> Rochester Institute of Technology

CHAPTER ONE

THE POWER OF PERSONALIZATION

WE HAVE ALL EXPERIENCED IT. In a crowded room filled with party chatter, you can concentrate on your own conversation with two colleagues and, without too much trouble, ignore what is going on around you. Then a voice within earshot, coming from a conversation in another corner of the room, mentions your name. Your attention, and perhaps your head, turns to the other conversations. Even though your name is not shouted or even called out, your attention moves away from your own companions and toward the other conversation. Rude behavior on your part, yes, but it is almost impossible for you to resist this distraction.

The ability to focus on one perceptual stimulus among many is called *selective attention*, and has been a mainstay of the study of cognitive psychology since the 1960s.[1] If we think of all the information that comes in through our senses at any given moment, we realize that it would overwhelm us if we weren't able to focus on one input stream at a time. Selecting one particular input stream enables us to pay close attention to what is happening in that stream. When some powerful, compelling input arrives in a different stream, it interrupts our attention long enough for us to analyze whether we want to turn to it. A name has this power to interrupt and grab our attention away from what we're doing. Other pertinent stimuli can distract us as well. For horse lovers, for example, a mention of the Kentucky Derby or a particular breed of horses will turn heads.

PERSONALIZATION DEFINED

How can marketers harness the attention-grabbing power of a customer's name or some other personal information in an advertising message? First we need to define what we mean by personalization and its cousin, customization. The term *personalization* describes a marketing communication tactic. Zahay and Griffin define personalized communication as "a specialized flow of interactive communication between parties," where a two-way communication is intended.[2] *Customization*, on the other hand, is most often associated with the product and the broader marketing strategy; products or services are designed for a clearly defined set of customers. The segmenting and targeting of groups of customers is the cornerstone of this marketing strategy.[3] Customizing a marketing program to a single customer is at the far end of the targeting continuum, where 1-to-1 communication with individual consumers provides the data to go back and design specialized products and services.

Customized products range from the creation of a single object for a particular customer, like a made-to-measure business suit, to the mass customization now seen in manufacturing.[4] An example of mass customization is the Dell Web site that allows a user to configure a new PC and determine how much it will cost. Sophisticated information technology (IT) at the point of purchase facilitates order entry, enables individual customers to design their own products, and gives them a choice of manufacturing and delivery options, all as simply accomplished as traditional mass production.

The term *customized marketing communications* has been used interchangeably with the term *personalization* over the last decade. What distinguishes personalization is the intent and use of the data.[5] Vesanen and Raulas reinforce this view by defining personalization as a business process. They catalogued a number of definitions of personalization used by prominent researchers in the marketing field, and from this work identified eight common elements within these perspectives: 1) the customer, 2) the dialogue with the customer, 3) gathering customer data, 4) analyzing customer data, 5) creating the customer profile, 6) customization, 7) marketing output, and 8) delivering marketing output. In fact, Vesanen and Raulas suggest that

> seeing personalization as a process makes it easier to understand what successful execution requires. The process of personalization contains the gathering and analyses of customer information from internal and external sources and customer interactions, customizing the marketing mix elements based on a customer profile, and targeting of marketing activities.[6]

When marketing managers create a personalized campaign, they analyze the available data to determine which potential customers will be most likely to respond to a uniquely crafted offer. Including the name of the recipient in a customized offer is not even necessary. For example, by data mining its transactions, a home flooring retailer learned that within 6 to 8 weeks of installing hardwood floors, homeowners are interested in buying area rugs. The savvy retailer will drop a postcard to these homeowners within this time interval to let them know of special savings on area rugs. The postcard need not acknowledge the recent installation of new flooring; it simply anticipates the homeowner's situation and announces the store's savings opportunity on area rugs.

PERSONALIZATION MEDIA OPTIONS

Personalization is defined as the complex range of marketing materials that can be designed using descriptive and behavioral information about individual customers and then delivered through a variety of media channels. The personalized message can be as simple as a specific name and address on a unique postal mailing or as complex as a unique promotional offer based on prior purchase history presented at the point of purchase.

Much of the excitement about personalized communications has centered on the Internet as the primary delivery medium,[7] through the use of customized Web pages or emails. These are very appealing options to deliver personalized messages to consumers. When users voluntarily visit a commercial Web site, they have elected to view content related to their interests. Many sites ask visitors to register and sign up for coupons or newsletters, yielding email addresses of viable "hot prospects" for the marketer. The use of email to deliver subsequent messages is appealing due to its near zero cost. However, email marketing has a problem: recent research summarized by the Direct Marketing Association (DMA) revealed that 67% of all email is spam.[8]

Some Web site personalization goes on "behind the scenes," as a function of computer programming algorithms that take the Internet user's search data and deliver links to related, mostly commercial, Web sites. The opportunity to fine-tune these algorithms is described in the following article.

PERSONALIZATION CATCHES THE USER'S EYE

by Gord Hotchkiss, *MediaPost*, September 13, 2007
Reprinted with permission of *MediaPost*

Last week, I looked at the impact the inclusion of graphics on the search results page might have on user behavior, based on our most recent eye tracking report. This week, we look at the impact that personalization might bring.

One of the biggest hurdles is that personalization, as currently implemented by Google, is a pretty tentative representation of what personalization will become. It only impacts a few listings on a few searches, and the signals driving personalization are limited at this point. Personalization is currently a test bed that Google is working on, but Sep Kamvar and his team have the full weight of Google behind them, so expect some significant advances in a hurry. In fact, my suspicion is that there's a lot being held in reserve by Google, waiting for user sensitivity around the privacy issue to lessen a bit. We didn't really expect to see the current flavor of personalization alter user behavior that much, because it's not really making that much of a difference on the relevancy of the results for most users.

But if we look forward a year or so, it's safe to assume that personalization would become a more powerful influencer of user behavior. So, for our test, we manually pushed the envelope of personalization a bit. We divided up the study into two separate sessions around one task (an unrestricted opportunity to find out more about the iPhone) and used the click data from the first session to help us personalize the data for the search experience in the second session. We used past sites visited to help us first of all determine what the intent of the user might be (research, looking for news, looking to buy) and secondly to tailor the personalized results to provide the natural next step in their online research. We showed these results in organic positions 3, 4 and 5 on the page, leaving base Google results in the top two organic spots so we could compare.

STRONGER SCENT

The results were quite interesting. In the nonpersonalized results pages, taken straight from Google (in signed out mode) we saw 18.91% of the time spent looking at the page happened in these three results, 20.57% of the eye fixations happened here, and 15% of the clicks were on Organic listings 3, 4 and 5. The majority of the activity was much further up the page, in the typical top heavy Golden Triangle configuration.

But on our personalized results, participants spent 40.4% of their time on these three results, 40.95% of the fixations were on them, and they captured a full 55.56% of the clicks. Obviously, from the user's point of view, we did a successful job of connecting intent and content with these listings, providing greater relevance and stronger information scent. We manually accomplished exactly what Google wants to do with the personalization algorithm.

SCANNING HEADING SOUTH

Something else happened that was quite interesting. Last week I shared how the inclusion of a graphic changed our "F" shaped scanning patterns into more of an "E" shape, with the middle arm of the "E" aligned with the graphic. We scan that first, and then scan above and below. When we created our personalized test results pages, we (being unaware of this behavioral variation at the time) coincidentally included a universal graphic result in the number 2 organic position, as this is what we were finding on Google.

When we combined the fact that users started scanning on the graphic, then looked above and below to see where they wanted to scan next with the greater relevance and information scent of the personalized results, we saw a very significant relocation of scanning activity, moving down from the top of the Golden Triangle.

One of the things that distinguished Google in our previous eye tracking comparisons with Yahoo and Microsoft was its success of keeping the majority of scanning activity high on the page, whether those top results were organic or sponsored.

Top of page relevance has been a religion at Google. More aggressive presentation of sponsored ads (Yahoo) or lower quality and relevance thresholds of those ads (Microsoft) meant that on these engines (at least as of early 2006) users scanned deeper and were more likely to move past the top of the page in their quest for the most relevant results. Google always kept scan activity high and to the left.

But ironically, as Google experiments with improving the organic results set, both through the inclusion of universal results and more personalization, their biggest challenge may be in making sure sponsored results aren't left in the dust. Top of page scanning is ideal user behavior that also happens to offer a big win for advertisers. As results pages are increasingly in flux, it will be important to ensure that scanning doesn't move too far from the upper left corner, at least as long as we still have a linear, 1 dimensional top to bottom list of results.

Gord Hotchkiss is the president of Enquiro, a search engine marketing firm. He loves to explore the strategic side of search and is a frequent speaker at Search Engine Strategies and Ad:Tech.

As hinted in this article, there is a limit to people's acceptance of highly personalized electronic marketing messages. For example, users of Web sites such as Facebook have rejected attempts to incorporate product advertising into their profiles or their email messages.[9] This type of intrusion into Internet users' personal content is viewed as a huge invasion of privacy.

Consumer protection groups are also voicing their concerns about the practice known as *behavioral advertising*.[10] As reported by Reuters, the Consumer Federation of America and the Consumers Union asked the FTC to create a *do-not-track* list that would prevent advertisers or Internet businesses to collect information about users without their explicit consent. This would include targeting advertising to a user based on Web searches or sites visited. Popular support for restricting such targeting activities was demonstrated in a March 2008 Harris Interactive survey with the finding that 59% of adult Internet users were "uncomfortable with Web sites using information about their online activities to target them with ads or content."[11]

While some electronic forms of personalization are being resisted, many printed forms are still generally acceptable. In research conducted by the Printing Industry Center at RIT, customers reported a preference for receiving U.S. postal mail from firms they already patronize.[12] Specifically, over three-quarters (82%) of the survey respondents liked getting catalogs from stores they frequent, and 67% also appreciated information about new products that they received from companies they patronize. In a 2006 study of 850 adults, InfoTrends asked respondents to identify their preferred method of receiving marketing messages. The majority, 61%, preferred direct mail. Only 21% chose email and even fewer, 6%, preferred telemarketing.[13] But in its 2008 study on the same topic with over 1,000 adults surveyed, InfoTrends found an increase in preference for email: 66.9% of the respondents indicated this was a preferred medium.[14] A cautionary note, however. The 2008 study used only an online survey, while the previous research also included mall intercept, face-to-face interviews. This methodology difference may account for a substantial portion of the increase in email preference.

In *Data-Driven Print*, we focused on *printed* personalized communications, including direct mail, catalogs, sales support materials created specifically for individual customers or dealers, and billing statements and other financial services information. In this book, we continue our emphasis on print, but place it within a direct marketing, multi-channel media context that includes Internet search and email marketing tactics. The continued focus is on printed communications for two reasons. First, new technology in digital color printing allows for cost-effective, customized printed promotional material with nearly unlimited design possibilities. Second, how print media will fare in the emerging world of electronic media is of keen interest to those who follow the economic viability of U.S. industries. Printed communications in the U.S.

still make up a significant part of the marketing communications budgets of many businesses. The figures for domestic advertising expenditures in 2006 tracked by Universal McCann reveal that nearly $60 billion was spent on direct mail.[15] This represents 21% of national spending on advertising media, up from $40 million in 1998. While much of the attention on personalized communications has focused on the Internet as the fastest-growing direct advertising medium, printed communications continue to make up a significant portion of U.S. marketing media expenditures.

While direct mail has grown, unsolicited or unwanted direct postal mail, or *junk mail* as it is commonly called, is wasteful on both personal and environmental levels and depletes advertising budgets on the commercial level. Unsolicited mail currently represents about half of the mail U.S. households receive.[16] Some estimate that as much as half of the mail delivered to a single postal address does not get opened, but instead goes directly from the mailbox to the wastebasket.[17] While overall direct mail response rates for direct order and fund-raising were 2.32% in 2006, other industry sectors show much lower rates. For example, the over five billion credit card offers that were mailed in the U.S. in 2004 produced a response rate of less than 0.5%.[18] While this low response rate apparently remained steady, at 0.5% in 2007,[19] the actual number of credit card mailings was expected to have decreased about 20% in 2008 to about 4.2 billion.[20]

Other challenges to the future of direct mail are concerns of privacy and data security. Consumers are increasingly voicing their choice to be removed from the mailing lists of firms they do not regularly patronize. As of May 2005, the Direct Marketing Association had 4,261,941 consumers on its Mail Preference Service.[21] Direct mail marketers are required to run their databases against this list and suppress the mail targeted to these individuals. As the number of "turned off" consumers grows, the long-term viability of the direct mail medium is in question. In addition, security breaches for financial services firms and information clearing houses such as ChoicePoint in 2005 have put nearly 50 million customers in the U.S. at risk.[22] Health information collected by insurance companies has also been hit by hackers. In April 2008, WellPoint, Inc. reported that over 130,000 customers may have had their data accessed by unauthorized users because of security lapses at their data storage vendor.[23] If savvy consumers refuse to share personal information with commercial firms because of these types of security concerns, personalization efforts will suffer from the loss of the information needed to make them work.

Unsolicited direct mail also poses an environmental problem. According to the EPA, 40% of America's trash is paper—71.6 million tons of it![24] The U.S. Postal Service (USPS) estimated that direct mail accounted for over 2 million tons of the volume of U.S. landfills in 2002.[25] While this is only a small proportion of the country's paper trash, direct mail is still vulnerable to criticisms from the environmental activists in the U.S. The USPS is mindful of these concerns and is developing "green" programs. For example, in May, 2008, the USPS targeted over 300,000 direct marketers with an advertising campaign that advised how to use more environmentally friendly materials.[26]

The premise of this book is that marketing firms can improve the results of their campaigns by eliminating the waste of unwanted advertising through the use of relevant, personalized promotional materials using the latest database and media technologies. Personally relevant marketing communications that are perceived as worthwhile by the receiver will reduce waste in both marketing budgets and environmental resources.

RATIONALE FOR THIS BOOK

Though the business case for personalized communications is strong, it requires the adoption of the latest information and media technology in order to implement sophisticated data-driven, content-rich, print and electronic media campaigns. As a starting point, the following case study from PODi, a clearinghouse for the successful implementation of personalized communications, illustrates the power of personalization. The Gannett Company leveraged internal data on newspaper subscriber transactions to trigger an automated print and email series of messages to nurture new subscribers and encourage them to renew. This case study clearly shows how far we've come in the sophistication of personalization efforts.

CASE STUDY: GANNETT CO. AUTOMATES MULTI-CHANNEL DIRECT MARKETING PROGRAM FOR NEWSPAPER RENEWALS

by PODi (Print on Demand Initiative)
Excerpted from *Best Practices in Digital Print 2007*
Reprinted by permission of PODi (www. podi.org)

Gannett Co., Inc, a major media company, wanted to find an inexpensive and automated way to bolster its new customer retention rates. A program consisting of personalized print and e-mail touches has helped achieve over 20% ROI for the company and has boosted 13-week retention rates an average of 13.7%.

PROGRAM OBJECTIVES

- Automate direct mail subscription retention efforts
- Reduce use of pre-printed materials
- Increase Web-based self-service and account management
- Increase auto-pay credit card and bank draft payments

SIGNIFICANT RESULTS REPORTED BY USER

- Over 20% ROI
- More customers providing e-mail addresses
- More customers converting to perpetual credit card payments
- Decreased customer service demands
- 13-week retention rates of markets in the program have increased an average of 13.7% year over year
- 26-week retention has averaged an increase of 7.6%.

DESCRIPTION

With 90 newspapers around the United States, Gannett was looking for an inexpensive yet effective way to centralize and automate some of its customer retention efforts. The solution the company created was a comprehensive system for direct marketing that the company dubbed Automated Direct Marketing (ADM). Tim Norris, Marketing Automation Specialist for Gannett, explains why the company decided to focus on retention efforts, "Retention marketing is critical to the growth of our newspapers but is often expensive and labor-intensive to execute." Working with markets both in large metropolitan areas as well as small community newspapers, ADM provides a centralized marketing process that automates the retention messaging for new customers.

CASE STUDY

ADM automates the production of personalized marketing messages via direct mail and e-mail. The system generates communications on-the-fly and printed materials are all produced in-plant in consolidated daily production runs to take full advantage of postal and production discounts. A company-wide e-mail vendor distributes the e-mail communications.

The retention curriculum is a series of nine print and e-mail touches targeted to new subscribers. The subscriber receives an order confirmation, start verification, guide to the newspaper, and service check versioned on their subscription payment type. Because of real-time automation, any change in a subscriber's status is reflected immediately in the messages. Each market works within a template but copy, images, prices, incentives, or premium and offer selections, even the timing the mail drops, are unique to each market.

All of the events (postcards and e-mails) are scheduled to be sent to a customer at specific times in their subscription cycle, so in essence most of the communications are scheduled. Some communications go to all new subscribers regardless of payment method at specific intervals. For example, every new subscriber receives a "Guide to the Newspaper" regardless of payment method.

The behavior of customer dictates which version of a postcard or e-mail they receive. For example, one message varies based on the payment method a customer is using. The paper asks someone who is not using EZ Pay, the automatic credit card payment plan, to convert to this plan. If a customer converts to EZ Pay between the "Thank You" piece, an e-mail they receive as soon as the transaction is complete, and the "Welcome" piece, an e-mail that is sent the first day the newspaper should start, then their Welcome piece will have different copy and will not contain a switch to EZ Pay incentive offer.

For postcard mailings not triggered by a specific event, each newspaper market can schedule when a particular touch is to take place. One market might want a service check, a postcard sent to a new customer to ask them to rate the service at 14 days into a new subscription. Another might choose a 21 day trigger. Norris says that they wanted to accommodate existing marketing preferences and allow markets to change production and mail dates.

The database framework that forms the underpinnings of the system is actually two separate databases, Norris explains. One is a large database that contains the transactional information for the entire newspaper chain. All the data is syncronized on a real-time basis, which allows the company to access data as needed for financial management or marketing efforts. The second database holds all the marketing-related information for the newspa-

CASE STUDY

pers such as logos, offers, newspaper names, addresses, etc.

Each night the system matches the marketing-related information to the transaction information and they use that combined data stream to print the variable data print communications. The e-mail provider receives the datastream in real-time so that the company can distribute the e-mail touches within minutes of the actual transaction occurring.

Developing the framework to marry the data from the two databases took several programmers 18 months to complete. This was a huge amount of custom programming, Norris explains. With more than 90 newspapers to support, ongoing maintenance on the system is also time consuming and includes implementing new versions of the software as well as generating management reports. Norris says that 67 markets are now using the program and that 5 more are in the roll-out stage. Initial results have helped encourage the

CASE STUDY

company to add more markets—the 13-week retention rates of markets in the program have increased an average of 13.7% year over year and 26-week retention has averaged an increase of 7.6%. Also, the program automation has decreased customer service time leaving time for more new customer acquisition efforts.

Wendy Hurwitz, Senior Director of Database Marketing for Gannett, also mentions some impressive results, "The ADM program has improved retention in participating markets and is posting an ROI in the 20% range. Online account management activity is up and more customers are providing email addresses. It has also increased the percentage of customers converting to perpetual credit card payments."

CLIENT	Gannett Co., Inc. http://www.gannett.com Gannett Co., Inc. is a leading international news and information company that publishes 90 daily newspapers in the USA, including USA TODAY. The company also owns nearly 1,000 non-daily publications in the USA and USA WEEKEND, a weekly newspaper magazine as well as newspapers in the United Kingdom and 23 television stations in the United States.
PRINTER	Inplant
HARDWARE	Two Xerox DocuColor 6060s
SOFTWARE	Developed in-house
TARGET AUDIENCE	New subscribers
DISTRIBUTION	1.2 million a year, printed daily
DATE	June 2005, ongoing
AWARD	In 2006 the ADM system helped win Gannett a Best in Show award in the Newspaper Association of America's ACME competition in the category of Database Techniques and Applications. The ADM Circulation Retention Program is also a finalist in INMA's 2007 Newspaper Marketing Awards Competetion in the direct mail category.

The remainder of this book expands on the why personalization is a powerful marketing tactic and how it can be implemented. We start in Chapter 2 by describing the nature of integrated media planning and what media have been popular choices for personalization. In 2002, RIT's Printing Industry Center started tracking the adoption and use of personalized marketing campaigns. In 2003, we found a relatively modest use of personalized marketing campaigns in the prior year, confirming the usually slow adoption patterns of any disruptive technology.[27] Chapter 3 reviews research published in 2007 and 2008, and benchmarks it against our 2003 data to determine whether personalization has increased or decreased in the intervening years. Chapter 3 also addresses the current use of personalization as a communications tactic, and the factors that facilitate and inhibit its widespread adoption. In 2003, we found two barriers to a more pervasive use of data-driven print: the lack of a top-level marketing strategy that benefits from a deep knowledge of customers, and the absence of the IT infrastructure required to execute such a program. New information in this edition documents current barriers and reveals a similar conclusion.

Chapter 4 addresses the need for businesses to understand how to develop a personalization strategy, both to acquire and retain customers. Chapter 5 examines the IT infrastructure requirements needed for successful implementation of these programs. Database fundamentals both for capturing customer information and for organizing it for creating personalized messages are discussed, as well as software requirements to create personalized print and electronic communications.

The latter part of the book examines other arenas for personalization. Chapter 6 describes how corporations are using these tactics for their own in-house communications. This chapter presents the view from the in-plant printer and shows how leveraging internal data can broaden the service offerings of this business function. A newly-minted term, *transpromo* (for "transpromotional"), describes the inclusion of marketing messages within mailed bills and statements.

Chapter 7 examines how businesses that have traditionally relied on the printed product can remain viable in the Internet age by providing marketing and other content in both print and electronic forms. Four cases studies are presented. Chapter 8 covers measurement issues and how to design feedback systems to track communication effectiveness. To get the most out of any marketing program, reliable feedback must first reveal whether the communication was received by the target audience and then measure its impact. Feedback is the key to determining financial gains and losses, as measured by return on investment (ROI) or cost per response (CPR), two simple metrics traditionally used by direct marketers. In addition to the mainstream media methods for assessing the impact of advertising, new media metrics are also reviewed. Chapter 9 summarizes the challenges of planning and implementing personalization efforts, and provides a starting point for any business that wants to create a dialogue with its customers.

NOTES

1. Neisser, *Cognitive Psychology*.
2. Zahay and Griffin, "Information Antecedents of Personalization and Customization in Business-to-Business Service Markets."
3. Kotler, *Marketing Management*.
4. Gilmore and Pine, "The Four Faces of Mass Customization."
5. McGrew, personal communication.
6. Vesanen and Raulas, "Building Bridges for Personalization: A Process Model for Marketing."
7. Godin, *Permission Marketing*.
8. DMA, *Statistical Fact Book*.
9. Creamer, "Think Different: The Web's Not a Place to Stick Your Ads."
10. Bartz, "Consumer Groups Urge 'Do Not Track' Registry."
11. eMarketer, "Is Behavioral Targeting Bothersome?"
12. Sorce, *Relationship Marketing Strategy*.
13. InfoTrends, *The Future of Mail 2006: Direct Mail, Transaction, and "Transpromotional" Documents*.
14. InfoTrends, *Trans Meets Promo… Is It More than Market Hype?*
15. DMA, *Statistical Fact Book*.
16. U.S. Postal Service, "The Household Diary Study: Mail Use & Attitudes in FY 2007."
17. Wann, "Affluenza: Curing the Bug. Five Ways to Fight Junk Marketing."
18. Campanelli, "Credit Card Solicitations at Highest in '04."
19. Synovate, "Mail Monitor, Synovate's Credit Card Direct Mail Tracking Service, Celebrates 20th Anniversary."
20. Synovate, "US Households Will Receive One Billion Fewer Credit Card Offers in 2008; Synovate Mail Monitor Shows Significant Decline in Offers to High Risk Consumers."
21. Direct Marketing Association (DMA), personal communication with the author, May 2005.

22. Dash, "Europe Zips Lips; U.S. Sells ZIPs."
23. Japsen, "Patient Data Faced Exposure: Wellpoint Client Records Left Open to Possibe Theft."
24. U.S. Environmental Protection Agency, "General Overview of What's In America's Trash."
25. U.S. Postal Service, "Lowering the Landfill Levels: Ad Mail's Environmental Impact is Small, Says New Study."
26. Syracuse, "You've Got Greenmail."
27. Mahajan, Muller, and Wind, Eds., *New-Product Diffusion Models*.

CHAPTER TWO

PERSONALIZED MARKETING COMMUNICATIONS IN THE INTEGRATED MEDIA WORLD

TODAY MORE THAN EVER, there are a variety of media channels from which to choose when deciding how to reach an audience. For example, if you are the manager for a local franchise hardware store, how will you allot your advertising budget? On local newspaper ads? Cable television ads? Direct mail to households within five miles of your store? Radio ads? By purchasing billboard space? By creating a promotional event involving hot air balloon rides in the parking lot? How does a business owner, marketing director, or media consultant working for a client choose among the media options available to contact prospective and current customers?

Even if you have made a decision about which media you'll use, you have to select among the multitude of media vehicles within each medium. For example, according to the FCC's 13th annual report to Congress on video competition, in 2006 there were 565 satellite-delivered national programming networks, compared to the 2005 total of 531 networks.[28] For magazines, the number of options is even larger. In 2007, the Oxbridge *National Directory of Magazines* counted 19,532 consumer magazines available in the U.S.[29] Businesses can also create their own advertising vehicles, such as catalogs. Oxbridge's *National Directory of Catalogs* listed 12,230 titles in 2008, a nearly 15% increase over 2004's total.[30]

The purpose of this chapter is to describe the advertising media landscape and how media decisions are made for the purpose of implementing a marketing campaign. Marketing executives (who create marketing programs) often outsource media decisions to an advertising agency, which in turn buys space or time in mass media outlets. In other cases, media decisions are made and implemented within the firm by the creator of the marketing strategy, usually the marketing manager or another senior executive.

This executive may also work directly with a print services provider or may produce the needed promotional materials with the firm's own internal resources. While the amount of media created by firms is difficult to measure, it is an important business process to consider. Let's begin our study of this topic by describing the nature of the media selection process within the context of marketing planning.

THE MEDIA PLANNING PROCESS

Media selection is but one of the steps in planning the marketing strategy of a firm.[31] As shown in Figure 2.1, a media plan starts with the marketing goals of a firm. For example, let's say a consumer products firm wants to introduce an extension to its shampoo line. The new product will repair sun-damaged hair for active young women aged 18-34. The marketing plan will be made up of four elements (often referred to as the four Ps):

- development of the *product* itself,
- determining its *price* and the budget for the marketing campaign,
- creating and delivering the *promotional* message, and
- "*placing*" the product for purchase by deciding what distribution channel(s) will be used.

All of these elements must be synchronized in order to create a successful plan. Two important elements of a marketing plan are the target market selection and identification of the marketing objective. In this example, let's say the product was developed to meet the needs of women who spend a lot of time in the sun and are worried about the effects of sun-damage to their hair. Let's say research has provided a demographic profile of this target audience: women 18-34 years of age who live in coastal states (e.g., Florida, California, and Texas).

Next comes the marketing objective that specifies a measurable outcome that will determine whether the plan is successful. In this case, since the consumer products firm is extending an existing shampoo line, the marketing objective may be, for example, to increase shampoo sales by 3% by the end of the year in specific geographic areas (e.g., Florida, California and Texas). To do this, the marketers establish a standard price and an introductory promotional price (for example, $2.79 as the standard price with $0.50 off during a trial period). In order to create an incentive for retailers to stock and place the item on the shelf, channel promotions such as a retailer promotional allowance for an end-of-aisle display may also be budgeted.

Once the marketing objective and the target market are established, the next step is to devise an advertising plan. A compelling message must be created and delivered to the target market using some advertising medium or a combination of media, all to be paid for within a predetermined budget. Most consumer products firms use advertising agencies for these complex tasks. Agencies develop the creative elements, recommend the media, implement the approved plan by producing the ads to be placed, and then buy space or time in the media selected.

The advertising planning process begins with specifying the promotional objective, which identifies the measurable response that is expected from the target audience after it has been exposed to the advertising. According to the standard hierarchy of effects,[32] most marketing communications planners realize that in order for a sale to result from advertising, consumers must:

- first, become aware of the product,
- next, be convinced that it will solve a problem they have,
- try it, and
- finally, perceive that it is the best option to solve the original problem, leading to a pattern of repeat purchasing.

In keeping with this approach, most promotional plans for new products identify awareness and trial as standard communications objectives. In the shampoo case, let's say the firm establishes the goal that 80% of the target audience would become aware of the product in six months' time. Once this objective is established, the message (content) of the advertising will be created. The choice of advertising media may come very late in the planning process, after the message has been determined and after the advertising budget has been allotted. Figure 2.1 summarizes this example.

With some exceptions, like IBM, many business-to-business (B2B) firms rarely use mass media to deliver advertising messages. B2B firms often have a smaller customer base (thousands rather than millions) and sell directly using their own sales force or regional value-added resellers, so the media choices for reaching their customers are less complex than for a consumer products firm. B2B marketing managers often create and implement promotional plans without the help of advertising agencies. However, a similar planning process is used. For these in-house marketing plans, the most frequently sought-after promotional objective is not awareness but lead-generation. A typical B2B marketing objective might be that 100 customers would inquire about the product or service within six months.

PLAN ELEMENTS

I. Marketing objective: Improve shampoo sales by 3% by the end of the year in coastal states (e.g., Florida, California, Texas)
II. Target market: women 18-34 years old who have sun-damaged hair
III. Marketing mix (Four Ps)
    a. Price: $2.79 with $0.50 off during trial period
    b. Product: Line extension with extra conditioners built-in
    c. Distribution channel: Place in current retail outlets
    d. Promotional plan:
        i. Objective: 80% of women aged 18–34 will be aware of the new product within six months
        ii. Channel promotion: Allowance for end-of-aisle display
        iii. Advertising
            1. Message: Impact of harmful effects of sun exposure are offset by using the new product
            2. Media options: Television, magazines, radio, direct mail
IV. Implementation – Tasks assigned to staff with timetables
V. Measuring results – Track sales of new product within target geographic areas

Figure 2.1 *Sample marketing plan for new shampoo line extension*

We have already seen that, in the marketing planning process, selecting the actual media to be used can be one of the last considerations the marketing manager or advertising agency will make. If a print services provider wants to get a piece of the marketing promotion budget, he or she must understand:

- who the decision makers are,
- what their role will be in the promotion planning process,
- the marketing objectives of the campaign, and
- how print will be able to deliver the firm's objectives of building awareness, generating leads, or getting an order in a cost-effective way.

As noted above, marketing executives can plan and implement advertising campaigns themselves or they can pay for the help of advertising agencies, media planners, or media services providers. This chapter examines the typical services provided by advertising agencies and documents trends in advertising media expenditures within the last few years. Next, it discusses the media decisions that marketing executives have historically been responsible for. Finally, it looks at current research on how printed advertising effectiveness compares with other media.

THE ROLE OF THE ADVERTISING AGENCY IN MEDIA SELECTION

In the early 1900s, advertising agencies served as the sales agents for the only mass medium at the time—print. Agencies bought space in bulk from magazines and newspapers and then sold it to clients, adding a 15% commission as the primary source of revenue.[33] This system remained in place in the U.S. until the 1990s when, following the lead set by European agencies, one of the oldest agencies spun off its media buying unit. Today's media specialist buys space and time (for broadcast) from the major media outlets and sells them directly to clients or through agencies to their clients, all for much less than the traditional 15% surcharge.

But media planning and buying is still a major service of advertising agencies today. In a 2003 Printing Industry Center study,[34] advertising agencies were asked to name the top three services that they provide (see Table 2.1). Media planning/buying was still the most common response, with nearly half of the agencies (48%) indicating it as a primary service. Creative development was the second most common response (43%), and customer relationship management was least common at 9%.

Table 2.1 *Primary services provided by advertising agencies*[35]

PRIMARY SERVICES OF AGENCY	PERCENT ANSWERED YES
MEDIA PLANNING/BUYING	48%
CREATIVE DEVELOPMENT	43%
GRAPHIC DESIGN	28%
SALES PROMOTION/COLLATERAL DEVELOPMENT	25%
PUBLIC RELATIONS	23%
BRAND CONSULTING	23%
DIRECT MARKETING	22%
DIGITAL BRANDING/WEB DEVELOPMENT	19%
CORPORATE IDENTITY	17%
OTHER SERVICES	12%
CUSTOMER RELATIONSHIP MANAGEMENT	9%

How does an advertising agency choose among the competing media? Traditionally, most advertising campaigns for consumer products began with a large broadcast television media purchase,[36] but today multi-channel and cross-media strategies are critical to successful business communications.

Table 2.2 shows how media expenditures for mainstream media changed in the five-year span from 2002 to 2007 and the resulting compounded annual growth rate (CAGR). The mostly positive growth rates for these media categories will likely be replaced by decreases in spending for 2008 and 2009, according to Bob Coen, director of forecasting at Magna Global, the central negotiation unit for Interpublic Group's media agencies.[37]

Table 2.2 U.S. ad expenditures by media type[38]

MEDIA TYPE	2002 GROSS EXPENDITURES ($ MILLIONS)	2007 GROSS EXPENDITURES ($ MILLIONS)	CAGR
TV - BROADCAST/CABLE	56,240	68,194	3.9%
NEWSPAPER - DAILY/WEEKLY	49,081	48,234	-0.4%
MAGAZINES - CONSUMERS/B2B	19,160	22,650	3.4%
RADIO	19,409	19,629	0.2%
YELLOW PAGES (PRINT)	14,810	14,981	0.2%
INTERNET	4,992	18,185	29.5%
OUT OF HOME	4,850	7,913	10.3%

Definitions:
TV = Broadcast (Network & Local) and Cable/Satellite (Network & Local)
Newspaper = Daily Print Advertising and Weekly Print Advertising
Magazines = Consumer Advertising and B2B Advertising
Radio = Broadcast Radio
Internet = Pure-Play Internet Advertising (advertising on websites not associated with traditional media companies; includes Google, Yahoo!, MSN, and eBay)
Out-of-Home = Traditional (e.g. billboards) and Alternative (e.g. marketing messages printed on valet parking tickets)

In selecting media for advertising, ad agencies have to balance a number of factors. A 2003 RIT Printing Industry Center study asked 250 advertising agency executives to indicate the top five factors that drive their media decisions for campaigns (see Table 2.3). The top three responses were consistent with good media planning processes—select the medium that reaches the right target audience at the right price to deliver the marketing strategy.

Table 2.3 *Factors driving the media choices in a campaign*[39]

FACTORS*	PERCENT ANSWERING YES
TARGET MARKET SELECTION OR DEMOGRAPHIC	72%
COST/BUDGET	63%
MARKETING STRATEGY	56%
PAST HISTORY OF SUCCESS	43%
CLIENT SPECIFICATIONS	35%
ROI TARGET	31%
CREATIVE FLEXIBILITY	23%
NEED FOR MEASUREMENT	19%
TIME AVAILABLE	15%
AVAILABILITY OF DATA/DATABASES	15%
NEED FOR A NEW LOOK	15%
NEED FOR PERSONALIZED MESSAGES	14%
* RESPONDENTS COULD INDICATE UP TO FIVE FACTORS.	

However, with the proliferation of media outlets and the problem of advertising clutter, advertisers are shifting away from major spending in broadcast media and leaning toward direct and interactive marketing methods, where the response to a marketing message can be captured easily.[40] Direct marketing methods, in general, are well positioned to deliver the accountability from marketing programs that clients are demanding in this economic climate. Whether it is a measure of lifetime value, return on investment, or cost-per-response, direct marketing has the advantage of a built-in measure of response. Table 2.4 shows the expenditures for "below the line" media (i.e., advertising media for which an ad agency does not charge a commission) from 2002 to 2007.[41]

Table 2.4 Below-the-line media expenditures, 2002, 2007[42]

MEDIA TYPE	2002 GROSS EXPENDITURES ($ MILLIONS)	2007 GROSS EXPENDITURES ($ MILLIONS)	CAGR
DIRECT MAIL	24,516	34,435	7.0%
CONSUMER PROMOTION	22,751	27,550	3.9%
CUSTOM PUBLISHING (OUTSOURCED)	2,255	4,739	16.0%
B2B PROMOTION PRODUCTS	15,627	19,553	4.6%
CATALOG	15,777	20,832	5.7%
TRADE SHOWS	8,524	10,964	5.2%
PR	2,575	4,271	10.6%

Definitions:
Consumer Promotion = POS and Coupons
Custom Publishing (Outsourced) = Magazines, Newspapers, and Tabloids

While the advertising agency can play a decisive role in selecting advertising media and executing ad campaigns, input from client firms is also critical. As shown in Table 2.3, client specifications are an important piece of the planning input (rated important by 35% of agencies). What are the latest challenges for marketing clients? When asked to define the driving force behind the pursuit of individualized market engagement, the top four factors reported by over 700 marketing executives surveyed by the CMO (Chief Marketing Officer) Council in 2008 were:[43]

- increasing retention and loyalty (39%),
- better use of marketing dollars and heightened ROI (37%),
- improving response and close rates (37%), and
- cutting through the clutter of competing messages (33%).

THE MARKETING MANAGER'S IMPACT ON DIRECT AND INTERACTIVE MEDIA SELECTION

Marketing executives are the primary decision makers for marketing campaigns that can be created in-house. In an InfoTrends study of direct mail marketers (primarily marketing managers), over half said that they were the sole decision makers in the design of their direct mail campaigns, with advertising agencies providing only an advisory role. Very few (11%) had given complete creative control to their agencies.[44] While we may think of printed direct mail as the most common direct advertising medium, direct marketing campaigns can be found in nearly every medium. Think about those late-night TV infomercials for some new exercise machine that will strengthen your back, enhance your chest, or reduce the size of your thighs. Table 2.5 gives the Direct Marketing Association (DMA)'s tally of how spending was distributed across all advertising media for direct response marketing campaigns in 2006.[45]

Table 2.5 *Direct response advertising expenditures in 2006*[46]

DIRECT RESPONSE MEDIA	DOLLARS SPENT IN 2006 (IN BILLIONS)
DIRECT MAIL, NOT INCLUDING CATALOGS	$32.9 B
DIRECT MAIL CATALOGS	$20.0 B
TELEPHONE	$45.7 B
INTERNET, NOT INCLUDING E-MAIL	$16.1 B
E-MAIL	$.4 B
DIRECT RESPONSE NEWSPAPER	$12.6 B
DIRECT RESPONSE TV	$21.7 B
DIRECT RESPONSE MAGAZINE	$8.5 B
DIRECT RESPONSE RADIO	$5.0 B
INSERT MEDIA	$.9 B
OTHER	$2.7 B
TOTAL	$166.5 B

Direct response marketers are leading the way in the use of multi-channel campaigns. For example, while the DMA found that, as of 2006, an estimated 61% of the revenues of businesses that use catalogs come from their paper catalogs, ALL catalogers also have a Web site or e-catalog, and 89% use email promotions.[47] In a study reported by Shop.org, almost half of the online retailers surveyed said that they also use printed catalogs, and that this would become a more important tactic in 2008.[48]

One of the reasons for the growth of these direct methods is the ability to personalize the marketing message, as is the case with telephone, Internet, and direct mail. In the CMO Council's 2008 study of over 700 marketing executives regarding their use of personalized marketing programs, email and letters topped the list of tactics being used.[49] Other results are also shown in Table 2.6.

Personalized Marketing Communications in the Integrated Media World

Table 2.6 *Top personalization techniques*[50]

PERSONALIZED MARKETING TACTIC OR TECHNOLOGY	PERCENTAGE
INDIVIDUALIZED E-MAILS AND LETTERS	66%
OPT-IN, PERMISSION-BASED MARKETING	47%
TARGETED DATABASE MARKETING LEVERAGING PERSONAL PROFILES	44%
PERSONALIZED E-MAIL PROMOTIONS BASED ON TIMING OF SIGN UP AND REGULAR INTERVALS THEREAFTER	40%
VARIABLE DATA PRINTING (VDP)	32%
PRINT ON-DEMAND COLLATERAL INCORPORATING PERSONALIZED CONTENT	31%
PERSONALIZED CARE AND HANDLING THROUGH CALL/CONTACT CENTERS	27%
PERSONALIZED URLS (INTERNET DOMAINS)	26%
CONSUMER-CUSTOMIZED WEB SITES (USER CONTROLS)	15%
WEB SITE PAGE DELIVERY BASED ON SEARCH HISTORY	11%
TRANSPROMO COMMUNICATIONS (PROMOTIONAL MESSAGES IN TRANSACTION DOCUMENTS SUCH AS PRINTED STATEMENTS)	9%

While the marketing executives in the CMO Council's study used a blend of electronic and printed media, they identified their primary future opportunities for personalizing customer communications as email (52%) and Web-based content (39%).[51] Printed personalization opportunities (e.g., sales literature at 14%) and new media (e.g., SMS text messaging) were regarded with much lower interest.

In sum, the statistics for the last five years of advertising media expenditures show a sharp rise in expenditures for Internet and email media. Further, in a world where marketing accountability is increasingly important, below-the-line advertising (direct mail, public relations, and promotional events) is growing at a faster rate than traditional broadcast advertising. Though the growth rates for printed direct marketing tactics are not going to keep pace with electronic forms, they still have a large absolute dollar value when added across all printed platforms such as direct mail, magazines, newsletters, and catalogs. Printed forms of advertising will retain their rightful place within the media mix if and only if they are found to be effective when compared with other media. The next section reviews the pertinent research on this topic.

EFFICACY OF PRINTED ADVERTISING

The impact of advertising has traditionally been evaluated using a variety of outcome measures such as aggregate sales for a brand, individual brand choice behavior, and the intermediate effects of awareness, beliefs and attitudes towards the advertised brand. A review of the standard metrics for assessing ad effectiveness across all media is presented in Chapter 8. Here, we review the findings from academic research, trade associations, and research consultants on how well print advertising performs when compared with other media. Many more studies are summarized on the Printing Industry Center's *Print-in-the-Mix* Web site at http://printinthemix.com.

Academic Research: Impact on Sales

The holy grail of an advertising effectiveness study is the ability of a researcher to unambiguously link a media user's exposure and attention to an advertisement and the subsequent buying behavior of that person. This research is difficult to carry out, and, when it does succeed, the results are often proprietary. What we find in the public record are correlational studies on indirect variables such as the amount of ad spending in one geographic location and the subsequent sales of the advertised product.

A study using this approach was conducted by Stafford, Lippold, and Sherron (2003), which tested whether there was a relationship between average weekly unit sales and the type of advertising used each week. For 121 weeks in 1999–2001, they analyzed pizza store sales data and the various advertising media used weekly by a particular franchiser in one metropolitan area. A regression analysis revealed that 63% of sales variability was explained by the media expenditure patterns. Further results of the regression indicated that the average weekly sales, without any advertising, were $10,696. The most powerful media combination was primary direct mail *and* national TV advertising, which produced a $1,057 increase in weekly sales.[52]

Vakratsas and Ma (2005) examined the effectiveness of media choice (magazine, network TV, and spot TV) on the monthly sales of competing brands of SUVs, factoring in *persistence effects*, or, the impact of advertising over time. They used a lagged regression analysis of monthly sales from the 1990s for Ford Explorers and Jeep Grand Cherokees with advertising expenditures as the predictor variables. For Ford, the portion of the total budget spent on the three advertising media studied was: magazines (36%), network TV (45%), and spot TV (12%). For Jeep, the numbers were: magazines (27%), network TV (30%), and spot TV (31%). The results showed that magazine advertising was more effective than network TV advertising for both brands in the long-term, and spot TV was the least effective advertising medium. Using a regression model to simulate the impact of changing the media mix on future sales, the authors predicted that increasing the advertising budget allocation for magazines could have improved the sales of both Explorers and Grand Cherokees.[53]

The majority of advertising effectiveness research published in academic journals describes the change in what people remember and feel about an ad after exposure, often in laboratory settings. Although this methodology does not directly measure the impact on buying, it does however assess the impact on what many believe is the necessary pre-condition to buying—a change in awareness, attitude, or emotional connection to an advertising brand. The research on these "intermediate effects" of advertising is reviewed in the next section.

Academic Research: Impact on Cognition and Affect

To test the intermediate effects of advertising on consumers' cognition, Sundar, Narayan, Obregon, and Uppal (1998) used a laboratory setting to assess the recall and recognition of an online text advertisement versus the same advertisement in a newspaper. The online and print ads were both presented in the context of news stories as one would see in a newspaper or online news site. While no difference in recollection appeared for the news stories in each medium, people who were exposed to the printed newspaper ad had higher recognition scores for its advertising content than those who received the online version.[54]

In measuring the affective impact of an ad, Calder and Malthouse (2004) of Northwestern University's Media Management Center created specific metrics to assess a reader's experience with printed media. They wanted to determine the media user's qualitative experience with magazines and whether this experience influenced the effectiveness of the advertising in it. Calder and Malthouse found that people regard time spent reading magazines as a luxury, and that they read magazines both for personal quiet time and to learn about new things.

The researchers' next step was to determine whether the readers' affective response to the medium itself impacted the message of an ad appearing in the medium. Using a mail survey, Calder and Malthouse asked a number of questions about an advertisement for a fictitious brand of bottled water that was printed in the magazine or newspaper sent to participants. They measured attitudes towards both the ad (with scales of "energetic," "soothing," etc.) and the qualitative experience of reading. The results indicated a positive relationship between the readers' affective responses to magazine reading itself and their attitudes towards the ad. That is, readers with a more positive response to survey statements such as, "I find this magazine to be a high-quality and sophisticated product," had more favorable attitudes towards the bottled water ad. The authors concluded:

> This research demonstrates that the way a person experiences a magazine or newspaper can affect the way the person reacts to advertising in the publication. For example, people who find the stories in a magazine more absorbing also have more positive reactions to the advertising in the magazine. Therefore, other things being equal, an advertisement in a magazine that absorbs its readers is worth more to the advertiser than the same ad in a magazine that does not absorb its readers as much.[55]

More recent research reinforces the impact of the audience's experiential responses to ads within a particular medium. Bronner and Neijens (2006) surveyed over 1,000 Dutch teens and adults about their advertising experience within chosen moments of media consumption. These experiences were classified as information, transformation, negative emotion, passing time, stimulation, and practical use. Print media (mail, free local papers, magazines, and newspapers) showed the highest average correlation between an advertising medium and a positive advertising experience. For example, if reading the newspaper was experienced as informative, the ads were found to be useful too.[56]

The impact of the advertising medium on the message, or *advertising/medium congruence*, has been discussed in earlier advertising research, though much of this work investigated the congruence between editorial content and the advertised product. In a recent study, however, *product/medium congruence* was tested for print advertisements on "creative" (unusual) media. Dahlén found that cognitive responses to ads (e.g., brand associations or attitudes towards an ad) were more positive for products advertised on congruent media than on incongruent media. For example, an insurance ad that appeared on an eggshell had higher ad credibility than the same ad appearing on an elevator panel. This result reinforces the conclusion that the medium itself can be a powerful contextual cue that impacts how ad messages are perceived.[57]

In the preceding studies published in academic journals, print media advertising consistently had a greater impact on market behavior (as measured by aggregate sales) and on intermediate effects (as measured by cognitive and affective responses to advertising) than did television or Internet advertising. These few studies emphasize the need for more research.

Next, we examine published research conducted by practitioners and industry firms, found in trade journals and consulting reports. Keep in mind that the following research studies did not go through the rigorous vetting process that is common to the academic review process.

Practitioner Research

Practitioner research can be categorized into two types, based on the sampled populations. The first type of practitioner research involves a self-selected population, such as a trade organization's survey of its membership of businesses, asking them to report the outcome of their advertising and media tactics on sales or inquiries. Research conducted by the DM A of its members is an example of this. The second type of practitioner research involves surveys of large samples of customers, media users, or buyers, measuring their media exposure (e.g., daily newspaper reading), buying behavior (past purchases), and/or their intended buying behavior (whether they expect to buy in the next few months). These studies are often commissioned by advertising organizations themselves, and the best studies use established, credible, unbiased research firms. First, let's examine research conducted by the DMA, regarding response rates to direct marketing methods.

The DMA's 2007 Response Rate Trends Report showed that its members obtained relatively high response rates across a variety of direct media. Table 2.7 reveals that, even though telemarketing had the highest average response rate as measured by direct orders, it also had higher costs per order—roughly double that of catalogs, and ten times more than email campaigns. Catalogs rated highest in lead-generating response rates, followed by email. However, email had a decidedly lower cost per lead (one fifth that of catalogs), making it a more efficient lead-generating medium. In the category of store traffic metrics, catalogs and direct mail were the top media.[58]

Table 2.7 *Average response rates by direct medium*[59]

CATALOG	RESPONSE RATE	COST PER ORDER/LEAD
DIRECT ORDERS & FUND-RAISING	2.24%	$28.27
LEAD GENERATION	6.12%	$26.00
TRAFFIC	10.34%	$3.65
DIRECT MAIL	**RESPONSE RATE**	**COST PER ORDER/LEAD**
DIRECT ORDERS & FUND-RAISING	2.15%	$19.90
LEAD GENERATION	1.61%	$27.38
TRAFFIC	5.35%	$6.84
INSERTS	**RESPONSE RATE**	**COST PER ORDER/LEAD**
DIRECT ORDERS & FUND-RAISING	0.11%	$26.62
LEAD GENERATION	0.57%	$20.06
TRAFFIC	0.12%	$34.50
E-MAIL	**RESPONSE RATE**	**COST PER ORDER/LEAD**
DIRECT ORDERS & FUND-RAISING	0.48%	$3.88
LEAD GENERATION	4.09%	$4.16
TRAFFIC	0.21%	$21.08
TELEMARKETING	**RESPONSE RATE**	**COST PER ORDER/LEAD**
DIRECT ORDERS & FUND-RAISING	2.53%	$43.70
LEAD GENERATION	2.46%	$67.97
TRAFFIC	N/A	N/A

The Magazine Publishers of America (MPA) does not survey its members. Rather, the organization works with the consulting firm Marketing Management Analytics (MMA) to compare the effects of changing the media mix for its clients (nearly 186 brands) over time. Using a proprietary measure of advertising effectiveness, MMA observed that, "When five percent of the media mix shifted from television to magazines, average media effectiveness more than doubled for the same brands."[60]

The second type of practitioner research assessing the impact of advertising media captures data from various end-consumers and media users. A study of this type was conducted in 2007, by the Newspaper Association of America (NAA), asking over 200,000 U.S. respondents what media provided useful information on bargains. Table 2.8 below shows that newspaper advertisements were regarded overwhelmingly as the leading source of information on deals and discounts.[61]

Table 2.8 *Media that provides useful information on bargains*[62]

CONSUMER TYPE	NEWSPA-PER ADS	TV ADS	RADIO ADS	MAGAZINE ADS	INTERNET ADS
ADULTS OF ALL AGES	66%	46%	39%	39%	24%
PRINCIPAL SHOPPERS	68%	47%	40%	41%	23%

A similar conclusion comes from research sponsored by Vertis Communications, a large direct mail and inserted media advertising consultant. Using a third-party research firm, Vertis annually surveys thousands of shoppers to determine what media they use to help them make shopping decisions. In the 2006 survey, 31% of participants chose ad inserts/circulars as the medium that most influenced their purchasing choices; television was chosen by 18% percent—the Internet by just 6%. For specific product categories, inserts/circulars were most influential in deciding where to shop for groceries, clothing, and home electronics. For these product categories, 55% of participants reported doing their shopping research using circulars and then making a trip to the store to purchase.[63]

Print advertising also works well as part of an integrated media mix. In 2008, a study commissioned by Google of just over 1,000 online search engine users, newspaper advertisements were found to influence their subsequent product research and purchase behavior. Among Internet-using newspaper readers, more than half (56%) stated that they had researched and/or purchased at least one product they had seen advertised in a newspaper in the previous month. Moreover, respondents ranked newspapers as more useful than the Internet to learn about promotions, and to decide where and when to buy.[64]

B2B Surveys

The media research we have reviewed thus far has concentrated on the household or citizen media user. As documented earlier in this chapter, over $10 billion per year is spent on communicating with business executives through trade magazines. In a 2007 B2B marketing mix survey conducted by Forrester Research for American Business Media (ABM), close to 65% of industry decision-makers read three or more industry-specific magazines per month, and 44% spend three hours or more per week doing so.[65] The top digital and mainstream media these decision-makers consult to "do their jobs" are presented in Tables 2.9 and 2.10.[66] Both online and printed industry-specific publications topped the list of media they rely on.

Table 2.9 Decision makers' top B2B digital media, 2007[67]

DIGITAL MEDIA	PERCENT USING
INDUSTRY-SPECIFIC MAGAZINE WEB SITES	70%
E-MAIL/ELECTRONIC NEWSLETTERS	70%
VENDOR WEB SITES	64%
WEB PORTALS (E.G., GOOGLE)	53%
WEB EVENTS	45%
SPECIALIZED BUSINESS WEB SITES	43%
GENERAL BUSINESS MAGAZINE WEB SITES	39%
B2B BLOGS	32%
SOCIAL NETWORKS	30%
MOBILE/WIRELESS	24%
RSS FEEDS	16%

Table 2.10 Decision makers' top B2B traditional media, 2007[68]

TRADITIONAL MEDIA	PERCENT USING
INDUSTRY-SPECIFIC MAGAZINES	69%
WORD-OF-MOUTH	67%
INDUSTRY-SPECIFIC CONFERENCES	65%
INDUSTRY-SPECIFIC TRADE SHOWS	62%
GENERAL BUSINESS MAGAZINES	55%
NEWSPAPERS	54%
DIRECT MAIL	51%
CUSTOM MEDIA	39%

CONCLUSION

In all of the studies reviewed in this section, printed advertising media were either superior or comparable to advertising in other media on a variety of outcome metrics. Magazine advertising was more effective than network TV advertising for promoting SUV brands over a 10-year period. Further, people who were exposed to printed newspaper advertising demonstrated a higher recognition of ad content than those who received an online version of the same message. For a food franchiser, the best sales resulted from the concurrent use of primary direct mail and national TV advertising. These results, considered in the context of the media spending patterns, reinforce the continued role for print in the media mix. The Google and ABM studies demonstrate that decision makers and the general public use a combination of media sources to research buying decisions and to do their jobs. Making advertising relevant is the key to its long-term acceptance and use by consumers. The next chapter will examine the place that personalized media have in the advertising mix.

NOTES

28. Federal Communications Commission, "FCC Adopts 13th Annual Report to Congress on Video Competition and Notice of Inquiry for the 14th Annual Report."
29. American Society of Magazine Editors, "Number of Magazine Titles, 1988-2007."
30. Marketing Charts, "Catalog Growth Powered by Religion, Auto, Education Categories."
31. Kelley and Jugenheimer, *Advertising Media Planning: A Brand Management Approach*.
32. McGuire, "An Information Processing Approach to Advertising Effectiveness."
33. Cappo, *The Future of Advertising*.
34. Pellow, Sorce, Frey, Olson, Moore, and Kirpichenko, *The Advertising Agency's Role in Marketing Communications Demand Creation*.
35. Ibid.
36. Cappo, *The Future of Advertising*.
37. "Spending in the Midst of 3-Year Drop, First Since Depression," 2008.
38. Veronis Suhler Stevenson, *VSS Communications Industry Forecast 2008-2012*.
39. Pellow, et al., *The Advertising Agency's Role in Marketing Communication Demand Creation*.
40. Schwartz, "Bigger and Bigger."
41. Veronis Suhler Stevenson, *VSS Communications Industry Forecast 2008-2012*.
42. Ibid.
43. CMO Council, "The Power of Personalization: The Impact and Influence of Individualized Content Delivery."
44. InfoTrends, *The Future of Mail 2006: Direct Mail, Transaction, and "Transpromotional" Documents*.
45. DMA, *Statistical Fact Book*, p. 223.

46. Ibid.
47. Ibid, pp. 64-66.
48. Magill, "Shop.org Projects Online Sales Will Rise 17% in 2008."
49. CMO Council, "The Power of Personalization: The Impact and Influence of Individualized Content Delivery."
50. Ibid.
51. Ibid.
52. Stafford, Lippold, and Sherron, "The Contribution of Direct Mail Advertising to Average Weekly Unit Sales."
53. Vakratsas and Ma, "A Look at Long Run Effectiveness of Multimedia Advertising and its Implications for Budget Allocation Decisions."
54. Sundar, Narayan, Obregon, and Uppal, "Does Web Advertising Work?"
55. Calder and Malthouse, "Qualitative Effects of Media on Advertising Effectiveness," p. 14.
56. Bronner and Neijens, "Audience Experiences of Media Context and Embedded Advertising: A Comparison of Eight Media."
57. Dahlén, "The Medium as a Contextual Cue: Effects of Creative Media Choice."
58. DMA, *2007 Response Rate Trends Report*.
59. Ibid.
60. Marketing Management Analytics, *Measuring Magazine Effectiveness: Quantifying Advertising and Magazine Impact on Sales*, p. 5.
61. Newspaper Association of America, "Newspapers: The Source for Bargains."
62. Ibid.
63. Marshall Marketing & Communications, "Vertis Customer Focus, Retail 2006: Media & Ad Inserts."
64. Newspaper Association of America. "NAA Analysis of New Google Research Finds Newspaper Advertising Drives Online Traffic, Consumer Purchasing."
65. Forrester Research, "The B2B Digital Marketing Shift."
66. Forrester Research, "The Digital Transformation."
67. Ibid.
68. Ibid.

CHAPTER THREE

THE CURRENT STATE OF PERSONALIZED MARKETING COMMUNICATIONS

IN DATA-DRIVEN PRINT, WE CONCLUDED THAT the current media planning environment could encourage the use of direct marketing campaigns because of its built-in capability to track response rates. New research reported in the previous chapter indicates that media decision makers are embracing interactive marketing methods, but that electronic media are the primary growth areas for personalized communication. This chapter reveals the extent to which personalized marketing communication is actually being delivered in campaigns, and specifically in print media. The growth of database technologies and an appreciation for 1-to-1 marketing tactics may fuel the demand for additional personalization activities.

Research conducted by RIT's Printing Industry Center in 2003 benchmarked the amount and complexity of personalized advertising in the U.S.[69] Two distinct populations—advertising agencies and marketing executives—were queried, because both play a role in media selection decisions. The "creatives," as they are sometimes called in advertising agencies, are important because of their influence on media selection decisions during campaign design and execution. While marketing executives are the final decision makers when working with agencies, they have a broader role in the design of below-the-line campaigns. For example, they may be the primary marketing strategists in the design and distribution of in-house marketing communications targeted to existing customers (e.g., a customer newsletter). The following section reviews more recent research and compares it with the Printing Industry Center's 2003 benchmark data, in order to determine how much progress has been made in the growth of personalized marketing communication.

AMOUNT AND COMPLEXITY OF PERSONALIZED COMMUNICATIONS

How many of today's marketing messages are personalized? We can start by reviewing the data from various industry trade associations such as the DMA. In its 2002 report on postal and email marketing, the DMA found that 62% of its members reported that they "sometimes" personalize information about a customer beyond name and address.[70] In the DMA's 2007 Statistical Fact Book, personalization is reported by 71% of the sample as a tactic used most frequently by direct marketers.[71] A study of direct marketers conducted by InfoTrends in 2008 reveals a similar estimate—68% of the 383 marketers sampled indicated that they personalized marketing campaigns to some extent.[72]

Figure 3.1 illustrates the amount of personalization used annually by advertising agencies and marketing executives, based on the Printing Industry Center's 2003 study.[73] We found that an average of 23% (standard deviation = 27.11, median = 10%) of the work completed by advertising agencies involved personalization, while about 33% of the marketing executives' campaigns involved personalization. Only 3% of marketing executives had not produced a single personalized campaign in the previous year, but nearly 20% of the advertising agencies we surveyed had not produced a personalized campaign in the previous year.

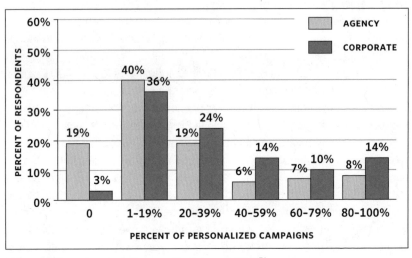

Figure 3.1 *Annual amount of personalization used, 2003*[74]

Recent research conducted by The Industry Measure research firm provides an updated look at this question. The firm periodically samples its panel of "creatives," composed of design and production employees in advertising agencies and graphic design firms. The firm's latest survey indicated that, in 2006, 27% of creatives reported that they had worked on a personalized, variable data job. These results cannot be directly compared to the 2003 Printing Industry Center data because the survey questions were very different. Industry Measure used a "yes/no" question, while the Printing Industry Center requested an estimate of the percentage of the campaigns that had been personalized in the prior year. However, Industry Measure asked a similar "yes/no" question in 2004, and found that 38% of the creatives surveyed had worked on such jobs in the previous 12 months.[75] This represents a substantial decrease of 11% from 2004 to 2006.

Industry Measure also assesses the use of personalization by publishers. In its 2006 survey, 15% of magazine publishers reported using personalized print.[76] While most personalized applications were for their own marketing materials, 58% of the magazine publishers reported using customized content for magazine covers.[77] An example of this effort was the Spring 2008 issue of the RIT student magazine *Reporter*, as shown in Figure 3.2 .

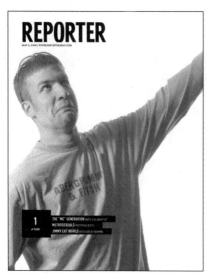

Figure 3.2 *Customized magazine cover example from Spring 2008 RIT Reporter*

The Current State of Personalized Marketing Communications

HELLO, YOU.
FROM THE MAY 2, 2008 EDITION OF *REPORTER*
WWW.REPORTERMAG.COM

You may not have realized it when you grabbed this off the stands, but the magazine you hold in your hands is the only copy in existence. Of the 10,000 magazines *Reporter* printed this week, no two are exactly alike. What you see before you is something truly unique; literally, it is one of a kind.

HOW DID WE DO IT?

On April 9, 11, and 14, Reporter set up stations in the lobbies of the SAU, Nathaniel Rochester Hall, and Building 70. At each of these stations, our photographers pulled aside passerby and shot their portraits on a white background. The only requirement was that they have an RIT ID card to scan, and that they give consent to use their portraits. In the 27 hours of shooting that were covered, several hundred portraits of RIT students, staff, and faculty were taken.

To handle the massive amounts of data, our Online Editor, Chris Zubak-Skees, wrote software to combine information from the scanned IDs and the photos from tethered capture of the portraits. The software put the images into folders sorted by University ID numbers.

After all the photos were shot, our Photo Editor, Dave Londres, went through each folder manually and deleted all but the best. He also cropped every image to fit in the available space and be a part of a cohesive design. The photos were then treated as variables to be automatically plugged into templates laid out by our design staff, under our Art Director, Jason Sfetko. The ensuing process was made possible by a technique known as variable data printing.

WHAT IS VARIABLE DATA PRINTING?

Variable data printing is a relatively new term for a redefined printing process which has only recently become practical with the rise of digital printing. Basically, images or text are a part of a database that feeds an algorithm plugging into a larger document. The end result is dynamic publishing of print material, or many different versions of the same thing.

For this issue of Reporter, a separate record was created for each magazine. They were fed into the Adobe InDesign document using XMPie. Our Production Manager, John Carew, programmed it such that there would be no repeated images on any issue. In total, there are 10,000 separate issues. On these, there are 431 cover images, and 200 faces sliced up and used to create the larger, composite face on page 20.

The pages of this magazine containing variable data (the front cover, the back covers, pages 2, 20, and 22) were printed at the RIT Printing HUB on the Xerox iGen3; all others were printed using the Goss Sunday 2000 at the Printing Applications Lab.

The bulk of personalization efforts are used to enhance marketing tactics. The 2008 survey from the CMO Council of over 700 marketing executives questioned their use of personalized communications, budget allocations, and barriers of further adoption.[78] The CMOs were asked to classify the extent to which their firms used personalized communications in customer acquisition and relationship development programs with a ranking of low, moderate or high. Only 17% indicated that their firms used a high amount of personalization, while 39% replied moderate, and 44% said low. In order to compare these results with the 2003 Printing Industry Center study, we reclassified our continuous scale into three categories. These new categories are presented in Table 3.1, and are compared to the 2008 CMO data.

Table 3.1 *Amount of personalization used by marketing executives, 2003 vs. 2008*[79]

CMO 2008 "TO WHAT EXTENT DOES YOUR COMPANY CURRENTLY USE PERSONALIZED COMMUNICATIONS IN ITS CUSTOMER ACQUISITION AND RELATIONSHIP DEVELOPMENT PROGRAMS?"		PRINTING INDUSTRY CENTER 2003 "PERCENTAGE OF CAMPAIGNS ANNUALLY THAT INVOLVE PERSONALIZATION"	
LOW	44%	0-19% OF CAMPAIGNS	39%
MODERATE	39%	20-59%	38%
HIGH	17%	60% OR MORE	24%

Using the above categories, we can conclude that the results are strikingly similar between the two studies across the five-year period.

Another way to track progress in the last five years is to assess how much complexity in these personalized messages was normal. There are many ways that a campaign can be personalized, and a variety of media to choose from. A personalized direct mail campaign can be produced simply by purchasing a list of prospective customers that fit a specific demographic or lifestyle profile, and then using traditional database marketing practices. In this case, the only data unique to the individual receiving the message will be his or her name, address, and salutation. More complex forms of customized marketing communications require specific data that capture the behavior of customers to produce unique messages that can vary their content in the text or graphics.

For the benchmark 2003 Printing Industry Center data, we asked our marketing executives and advertising agencies to distribute 100 points among five levels of complexity to indicate how often they used each level. These levels of personalized advertising complexity are defined in Table 3.2. Both of our respondent groups indicated that nearly half of their marketing campaigns used the lowest level of complexity—the mail-merge option, which includes only a variable address and/or salutation. Only 27% of the advertising agencies and 18% of the marketing executives used graphics in their customized messages.

Table 3.2 *Levels of complexity used in personalized campaigns*[80]

LEVEL	ADVERTISING AGENCIES	MARKETING EXECUTIVES
VARIABLE ADDRESS AND/OR SALUTATION	46%	50%
VARIABLE ADDRESS AND/OR NUMERICAL INFORMATION IN FIXED FIELDS	15%	19%
VARIABLE ADDRESS, TEXT, AND NUMERICAL INFORMATION IN DYNAMIC FIELDS	12%	14%
VARIABLE TEXT OR NUMBERS AND GRAPHICS	14%	10%
VARIABLE TEXT OR NUMBERS AND VARIABLE GRAPHICS	13%	8%

Comparable data can be found in the 2006 InfoTrends study on the future of mail.[81] InfoTrends asked over 200 direct mail document business owners about the complexity of the personalization used in their direct mail campaigns. If we assume that the direct mail document business owners are marketing executives, we can use the InfoTrends data to determine whether any changes have been observed in the complexity of personalization used in marketing campaigns. InfoTrends found that 41% of the direct mail volume was not personalized at all (except for name and address), 27% was "limited," and 32% was highly personalized. Consistent with the 2003 benchmark data from the Printing Industry Center, the most frequent type of personalized communications were the simplest. In its 2006 report on the future of mail, InfoTrends commented on the changes in personalization rates its research had discovered from 2003 to 2006, and concluded that:

> There has been a shift since the last study, with not personalized increasing from 38% to 41%, highly personalized essentially flat and limited personalization dropping from 31% to 27%.[82]

While the InfoTrends and Printing Industry Center research results differed by the nature of the responses (three categories versus five) and the type of tactics queried (direct mail versus all campaigns), we can conclude that personalized marketing campaigns have maintained about the same degree of complexity over this time period.

The most recent Industry Measure research study also gave some insight into changes in the complexity of the personalized communications that advertising agencies have created. When asked whether they had worked on a specific type of personalized job (multiple responses were accepted), 49% of respondents had created a simple mail/merge job, 38% had created a full-color job with variable images and text, 21% had used pre-printed offset shells, and 13% had worked on a versioned piece in 2006. Compared to Industry Measure's findings from 2004, work on all but one category of complexity had declined. The mail-merge category had the only increase, with a jump from 46% to 49%. The incidence of the most complex type of jobs, incorporating both full-color text and images, decreased from 49% in 2004 to 38% in 2006.[83]

In sum, the most recent research indicates that we are at a stand-still (as of 2007) in the adoption of personalized communications when compared to 2003 levels. Whether we examine the amount or complexity of personalization, or differences between the agency and marketing executive, we see little evidence of an increasing trend. What accounts for this lack of growth?

FACTORS IMPACTING THE LEVEL OF PERSONALIZATION

In *Data-Driven Print*, awareness was a key barrier to the adoption of personalized printing technologies for use in creating direct mail. Consistent with the technology adoption model, a company must first be aware of new technology before it can try or adopt it.[84] As shown in Table 3.3, nearly two thirds of our research study groups were aware of the new printing technologies used for personalization in 2003. However, while over half of the advertising agencies had shown samples to their corporate clients, only 36% of the marketing executives reported that they had been shown samples by their agencies. The marketing executives' lower levels of exposure to samples of new print technology may be explained in part by the percentage that had been using an advertising agency the year before the study was conducted. (We had found that only 53% of the marketing executives in our sample had used an advertising agency during the prior year.)

Table 3.3 *Awareness of personalization print technologies*[85]

AWARENESS STATEMENT	ADVERTISING AGENCIES*	MARKETING EXECUTIVES*
I AM AWARE OF NEW PRINT TECHNOLOGY FOR PERSONALIZATION.	64%	62%
I HAVE (BEEN) SHOWN SAMPLES OF PRINTED COMMUNICATIONS USING THESE NEW TECHNOLOGIES.	57%	36%
* The respondents used a 5-point scale, where "5" means "strongly agree." The percentage reported is for "agree" ratings of 4 or 5 combined.		

Awareness does not seem to be an impediment today. In the 2008 CMO Council study, all respondents were either planning to deploy a personalized campaign (17%) or had used personalization before (39% had used personalization for 1–3 years and 44% had used it for 3 or more years).[86]

The perceived benefits of using personalization are apparent to marketing decision makers and influencers. When Industry Measure asked the creatives about the primary benefits of personalization in 2006, 41% agreed that it produced higher response rates, 33% said it could improve customer satisfaction, and 31% thought it could lower total costs.[87] In the CMO Council study, the top benefit of personalized communication was perceived to be that it "makes offers more relevant and meaningful to the prospect" (53%); "builds [a] closer, more intimate relationship with the customer" (48%); "increases campaign effectiveness and yield" (40%); and "increases the company's overall marketing effectiveness" (38%).[88]

By 2008, the benefits of personalization were well-known. So, what was the obstacle? Advertising agencies surveyed by the Printing Industry Center for its 2003 study had been asked why they did not recommend more personalized campaigns. The most common obstacle keeping these respondents from recommending personalization strategies to their clients was price, at 28%, with lack of a suitable database following at 23%. Nearly half of the respondents stated that "some other obstacle" kept them from recommending personalization to their clients. Of this group, nearly one-third wrote in that there was a lack of need for this type of advertising strategy. The marketing executive respondents in the 2003 Printing Industry Center research were also asked about obstacles. Lack of resources (money, databases, people, or knowledge) and lack of need were the top barriers mentioned. These obstacles remained the top inhibitors for both agency and creative professionals in 2006. Industry Measure reported the top barriers to personalized, customized variable data jobs were marketing savvy (presumably, the lack thereof, at 28%), database quality (28%), technical skills (24%), and lack of demand from clients (21%). Cost was reported by 16% as a barrier, which was a decrease from the 2004 Industry Measure study of creatives.[89]

Nearly half of the marketers interviewed by the CMO Council indicated that the key challenges for integrating personalized communications into their marketing programs were inadequate infrastructure (49%), lack of customer data and insight (46%), and cost and complexity (43%). Regarding the "people side" of the business, 27% of the CMOs in the study indicated that they lacked internal competency, and 20% stated that "management mind set" was a barrier.[90]

When asked about the adequacy of their knowledge of their customers, the majority of the CMOs reported that it was either "good" or "fair" (74% in total), but only 13% indicated that they had a comprehensive profile of their customers that included transaction data.[91] This is where the gap is revealed between what marketers want to do and what they can do. When they were asked about the most important data they used in designing a personalization program, nearly half of the respondents in the CMO Council study (47%) indicated that it was purchase history. Since only 13% included this information in consumer profiles, we can only imagine the frustration these marketing decision makers must have experienced. Other important data used in personalization programs (reported by over 40% each) were profitability and length of the relationship, customer classification (business or individual), and geographic location. Customer demographics and lifestyles were identified as the most important factor in personalization programs by no more than one quarter of the respondents in the CMO sample.[92]

There is a definite link between the internal data capabilities of a firm and the use of more complex forms of personalization. Further analysis on the Printing Industry Center 2003 data tested whether those firms with more data management and analysis capabilities also used more complex forms of personalized messages in their marketing communications programs. The results revealed that firms that had CRM (customer relationship management) systems and used data mining and campaign management software also used more complex forms of personalization.[93] Table 3.4 compares the respondents' use of CRM, data mining, and campaign management tools to the percent and complexity of personalized campaigns that were mounted.

Table 3.4 *Database functions and level of personalization used*[94]

LEVEL OF PERSONALIZATION USED	USE OF CRM		USE OF DATA MINING		USE OF CAMPAIGN MANAGEMENT TOOLS	
	YES	NO	YES	NO	YES	NO
PERCENT OF CAMPAIGNS PERSONALIZED	35.2%	34.9%	32.2%	34.4%	32.3%	34.2%
LEVEL OF COMPLEXITY						
VARIABLE ADDRESS AND/OR SALUTATION	45.3%	50.3%	46.5%	50.8%	44.7%	51.3%
VARIABLE ADDRESS AND/OR NUMERICAL INFORMATION IN FIXED FIELDS	27.6%*	15.1%*	25.8%*	14.7%*	20.0%	18.4%
VARIABLE ADDRESS, TEXT, AND NUMERICAL INFORMATION IN DYNAMIC FIELDS	16.1%	12.9%	12.9%	15.2%	22.1%*	10.5%*
VARIABLE TEXT OR NUMBERS AND GRAPHICS	9.2%	11.8%	9.7%	10.9%	8.6%	11.2%
VARIABLE TEXT OR NUMBERS AND VARIABLE GRAPHICS	4.2%	9.9%	6.8%	8.2%	6.3%	8.4%
N	33	79	48	78	41	85
* significant at p<.01, anova						

The amount of personalized campaigns did not vary in relation to any of the database technology applications. That is, whether firms had a CRM system or not, or did or did not use data mining or campaign management, the average amount of campaigns that were personalized still ranged from 32% to 35%. However, the complexity of the personalization was impacted by the data resources of the firm. Specifically, second level complexity, where variable address and/or numerical information are placed into fixed fields, was used more by firms that had CRM systems than by those that did not (27.6% and 15.1% respectively, p = 0.01). This level of complexity was also used more by those possessing data mining capabilities (25.8%) than those that did not

(14.7%, p = 0.01). The middle level of complexity, where variable address, text, and/or numerical information are placed into dynamic fields, was used more frequently by those with campaign management tools (22.1%) than those without (10.5%, p = 0.01). There were no differences in the use of the two highest levels of complexity in relation to the presence of database technology systems in the firm.

We conducted another test to determine whether the frequent use of personalization was related to more complex forms of personalization. A correlation analysis was performed to test the number of personalized campaigns implemented and the complexity level of the personalization. As shown in Table 3.5, those firms that conducted more personalized campaigns reported a more frequent use of the highest level of customization ($r = .214$, $p < 0.05$).

Table 3.5 Correlations with level of complexity[95]

LEVEL OF PERSONALIZATION USED	PERCENT OF CAMPAIGNS PERSONALIZED	SIZE OF FIRM	PERCENT OF MEDIA BUDGET SPEND ON DIRECT MAIL
PERCENT OF CAMPAIGNS PERSONALIZED	—	−.181*	.207*
LEVEL OF COMPLEXITY			
VARIABLE ADDRESS AND/OR SALUTATION	−.150	.048	.009
VARIABLE ADDRESS AND/OR NUMERICAL INFORMATION IN FIXED FIELDS	−.053	−.085	−.003
VARIABLE ADDRESS, TEXT, AND NUMERICAL INFORMATION IN DYNAMIC FIELDS	.064	.131	.153
VARIABLE TEXT OR NUMBERS AND GRAPHICS	.095	.026	−.112
VARIABLE TEXT OR NUMBERS AND VARIABLE GRAPHICS	.214*	.041	−.146
N	148	111	163

* significant at p<.05

Table 3.5 also presents the correlation between firm demographics (the size of the firm and the percent of the media budget spent on direct mail) and the amount and degree of personalization used. There were two significant correlations discovered. More personalized campaigns were used by smaller firms (r = -.181, p<.05) and by firms with a larger percentage of their media budget spent on direct mail (r = .207, p<.05).

We also tested whether the nature of the database applications varied by industry. Financial services firms used significantly more data mining (48%) and campaign management tools (46%) than other types of firms. When broken down by firm size, larger firms were more likely to use data mining (39%) than smaller firms (21%). Firms with larger advertising budgets ($1 million and over) used data mining, campaign management, and data cleansing software significantly more than those with smaller budgets. However, there were no differences in whether a firm had a CRM system by size of marketing budget or industry.

It should be no surprise that a firm's resources are deployed in marketing programs. The InfoTrends 2006 study analyzed what kinds of firms used digital color in their direct mail programs. They found that:

> Detailed personalization is also strongly correlated with the use of digital color; 42% of the mail sent by those who produce digital color direct mail is highly personalized, while half of the mail sent by those who do not print digital color is not personalized at all.[96]

But technology is not the silver bullet. As we noted in Chapter 2, the growth of Internet search advertising gives any business a low barrier to enter this new world of personalized direct communications with potential customers. The article below shows that, even with electronic forms of personalization, what is said and how it is delivered matter.

**PERSONALIZATION PITFALLS OF INTERNET SEARCH
IRRELEVANT ADS BREED TURNED-OFF CONSUMERS**

by David Szetela, MediaPost Search Insider, Thursday, March 20, 2008
Reprinted with permission of MediaPost

In a recent opinion piece in The Wall Street Journal, Esther Dyson described the growing irrelevance of traditional online ads, and the growing importance of advertising on social networks. She contended that Internet users have become inattentive to ads due to their lack of relevance to site visitors (so-called "banner fatigue"). She also predicted that someday (soon?) people will be able to "friend" advertisers and only see their ads — or offers tailored to the information users have chosen to supply — on pages of specific sites they visit.

I couldn't agree with Dyson more regarding the irrelevance of ads. In my weekly Search Engine Watch column, I've repeatedly underscored the fact that everyone loses — site visitor, advertiser and publisher — when displayed ads aren't relevant to the page content, and hence to the site visitor. The visitor breezes right by the ad (or worse, is turned off by the severity of its irrelevance), the advertiser doesn't get the click, and the site publisher doesn't get the revenue that Google or Yahoo would have paid.

Here's an example of how bad the situation's become: one of our clients is a book author whose first novel was recently reviewed in the New York Times Sunday book section. Since the Times serves ads in non-premium positions (like the bottom of the page) using Google's self-serve AdSense system, I tried to use Google AdWords to display ads on the online book section pages (as well as other pages on the Times site). Result: No dice. No matter how high I placed my click bids (within a reasonable range), my ads wouldn't appear on the page. Why? Most likely because other advertisers were bidding higher.

I noticed that each of the book section pages displayed ads from a prominent mortgage company, a well-known computer manufacturer, and a ringtone vendor. None of which was relevant in any way to the content of the page, much less in any direct way to book readers in general.

As a trend, nobody is served well by this phenomenon. Advertisers get fewer responses when content and ad/offer don't match. Publishers get less revenue when ads are paid for on a CPC basis — which is becoming more prevalent, especially with the meteoric growth of the Google AdWords network. And site visitors lose because the ads have become a negative experience and at best are ignored and at worst an annoying distraction, like an ad for a massage parlor in a religious magazine.

While I agree with Dyson that the future looks bleak for online contextual advertising, I think the various ad-serving players can and will wise up and incorporate technology and policies that will reverse the trend. Smaller contextual networks like ADSDAQ claim to use algorithms that more closely match page content with relevant ads. Google AdWords recently rolled out a tool that lets advertisers choose to exclude their ads from whole categories and topics of sites that they deem irrelevant.

In the premium ad space segment, manual review and more-stringent publisher policies may be required to ensure irrelevant ads aren't served. This is a difficult function to express algorithmically and build into ad service software - but let's never underestimate the ingenuity of properly motivated software developers humming with Red Bull, and the creative energy that's humming through the online advertising industry.

CONCLUSION

There is significant discussion in marketing circles about customized marketing communications and 1-to-1 marketing. Consultants tout the value of reaching the target audience with right-place and right-time exactness. One-to-one marketing, they claim, allows clients to use knowledge about customers' preferences to rise above the din of competitive messages. Even though the 1-to-1 revolution was well underway in 2003, neither the advertising agencies nor their corporate clients had broadly embraced marketing campaigns designed around these principles. Sixty-four percent of the advertising agency respondents in our 2003 study agreed that they were aware of the technology, and 57% reported that they had shown samples of these campaigns to clients—but only 23% of the work they created used variable data. Moreover, 46% of the advertising agencies' variable-data jobs and half of the marketing executives' jobs involved only a simple mail-merge.

How much progress has been made since then? The most recent research indicates that we have not made much progress. Examining either the amount or complexity of the personalization for both agency creatives and marketing executives shows that there is little evidence of an increasing trend. However, there is evidence that the decision makers see the advantages of personalization and are willing to use it. The inadequacy of customer data within organizations continues to be the principal barrier in the deployment of personalized campaigns. Firms with more information resources used more complex forms of personalization. More personalized campaigns were used by smaller firms and those with a history of using direct methods.

The next chapters provide insights into overcoming the database barrier. But we need to make an important point—simple personalized campaigns do not need to be simple-minded. The design of a direct mail piece itself may not be a good indicator of the intelligence that was used to generate it. For example, a black-and-white postcard sent in a timely manner to customers who are predicted to buy at certain intervals may be a highly successful tactic for retailers. The use of sophisticated predictive analytics does not require the use of color and graphic images in a mailing. And, given its lower cost, a simple, black-and-white design might even produce greater ROI than a more expensively-produced mailer. So, successful personalization efforts can be implemented by utilizing some basic database marketing strategies.

NOTES

69. Sorce and Pellow, *Demand for Customized Communications by Advertising Agencies and Marketing Executives.*
70. Direct Marketing Association, *The State of Postal and E-Mail Marketing: New List Trends and Results.*
71. Direct Marketing Association, *Statistical Fact Book.*
72. InfoTrends, *Trans Meets Promo.... Is it More than Market Hype?*
73. Sorce and Pellow, *Demand for Customized Communications by Advertising Agencies and Marketing Executives.*
74. Ibid.
75. Industry Measure, "Variable Data Printing/1:1 Personalization: 2007."
76. Ibid.
77. Ibid.
78. CMO Council, "The Power of Personalization: The Impact and Influence of Individualized Content Delivery."
79. Ibid.
80. Sorce and Pellow, *Demand for Customized Communications by Advertising Agencies and Marketing Executives.*
81. InfoTrends, *The Future of Mail 2006: Direct Mail, Transaction, and "Transpromotional" Documents.*
82. Ibid, p. 22.
83. Industry Measure, "Variable Data Printing/1:1 Personalization: 2007."
84. Rodgers and Chen, "Post-Adoption Attitudes to Advertising on the Internet."
85. Sorce and Pellow, *Demand for Customized Communications by Advertising Agencies and Marketing Executives.*
86. CMO Council, "The Power of Personalization: The Impact and Influence of Individualized Content Delivery."
87. Industry Measure, "Variable Data Printing/1:1 Personalization: 2007."
88. CMO Council, "The Power of Personalization: The Impact and Influence of Individualized Content Delivery."
89. Industry Measure, "Variable Data Printing/1:1 Personalization: 2007."
90. CMO Council, "The Power of Personalization: The Impact and Influence of Individualized Content Delivery."
91. Ibid.
92. Ibid.
93. Sorce and Pellow, "Amount and Complexity of Personalization in Marketing Campaigns."
94. Ibid.

95. Ibid.
96. InfoTrends, *The Future of Mail 2006: Direct Mail, Transaction, and "Transpromotional" Documents*, p. 189.

CHAPTER FOUR

PERSONALIZATION STRATEGIES FOR CUSTOMER DEVELOPMENT

THE MOST PERSISTENT BARRIER to the widespread adoption of personalization is an inadequate information infrastructure, without which there can be no actionable customer data and insight. Having a well-designed customer data management system is a critical first step. As illustrated in Figure 4.1, reprinted from an article by Payne and Frow (2005), an information infrastructure is developed by first understanding a firm's business strategy and then obtaining a deep knowledge of customers.[97] These inputs articulate the value proposition, or the set of benefits the firm can offer to its customers to satisfy their needs.[98] The value proposition will determine the messages and media used in the multichannel integration process. Once an information system is in place to gather and store data—these systems are often referred to as customer relationship management (CRM) systems—data mining algorithms can be used to gain insight into customer buying behavior that will be the basis of new marketing campaigns.

GAINING CUSTOMER INSIGHTS

The basic requirement of a successful marketing strategy is having a deep knowledge of the firm's customers. Based on recent observations by advertising consultants Briggs and Stuart (2006), some of the most respected firms do not know enough about their customers and the communication strategies that influence their buying behavior. Because of this, Briggs and Stuart claim that over one-third of all advertising spending is wasted. Briggs and Stuart posit that the starting point is to understand why people buy a product. More specifically, why do customers choose one brand over another?

Obtaining insights into consumer behavior starts with the knowledge of current customers, gained from a thorough analysis of the transactions they

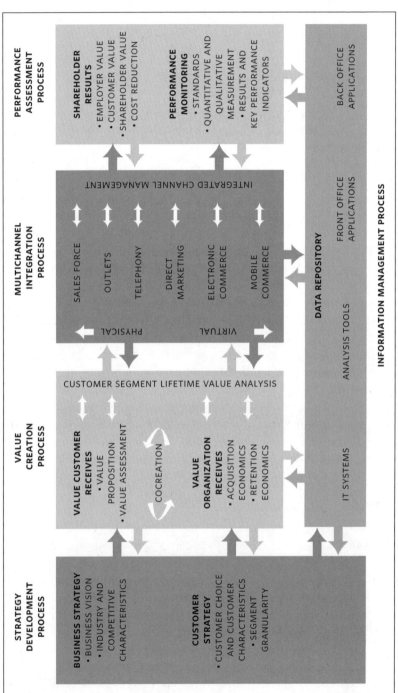

Figure 4.1 A conceptual framework for CRM strategy[99]

have with a firm. Most firms have some kind of customer database. But, as noted in the previous chapter, nearly one-half of marketers claim that they don't have the right data or the ability to gain insights into consumer behavior from their existing database. What, then, should the customer database contain?

It would be optimal to have a purchase history and a demographic profile of each customer in the database. While this sounds simple, for many purchase situations (such as for frequently-purchased consumer products), the manufacturer does not have access to information about the individuals who buy through retail outlets. Even the retailer is challenged to obtain this information, as more and more customers are becoming wary of sharing personal information. This explains the proliferation of loyalty programs at the retail level, where customers are willing to share information during a registration process for a loyalty card in exchange for discounts and other shopping privileges.

To get more information about why customers shop at a given store or buy a particular brand, marketers must go beyond simple descriptive and transactional information. Additional in-depth analysis of current customers can also provide insights into the lifestyle "trigger events" that bring customers into the marketplace to solve problems. Clearly, events such as buying a new home or a child going off to college trigger a number of needs for new types of products. Identifying trigger events not only allows a firm to predict when a current customer is open to learning about new products (for the purposes of cross-selling and upselling), but also provides a method to help identify prospective customers who are in similar situations.

Additional information about a customer's perception of a firm's products and services can come from the increasingly-important, post-sale interaction that a firm has with the customer at its customer service center. Telephone call centers, Web sites, and blogs can be vital sources of information of how customers use, misuse, or misunderstand products and services. These manufacturer-created systems are critical for a consumer products firm whose products are distributed through self-service retailers that do not provide highly knowledgeable sales personnel to assist customers in product choice or feature explanation. Service centers not only assist customers on how to use a feature-rich product, but also provide important feedback to the product design teams for the next round of continuous improvement. Success in helping customers, whether by clarifying billing problems or explaining how to use a new feature, is critically important in building customer satisfaction with a firm and its products or services.

All of the information from these customer touch points should be located in a centralized repository or data warehouse, as shown in Figure 4.1. Keeping track of customers one at a time is necessary to leverage their information to create the 1-to-1, personalized marketing messages illustrated in this CRM model. While customer retention is a top marketing priority, much of the investment in advertising is focused on acquiring new customers. In this case, marketers take the insights gathered from their current customers and apply them to the remaining target market.

As an example, the analysis of the lifetime value (LTV) of a banking customer may reveal that there are five major events that trigger the need for financial services: marriage, purchase of the first home, birth of the first child, a child going to college, and retirement. Identifying people or households who are at the thresholds of these events can be the designated marketing effort for the bank in order to find new customers. Though not much may be known about new prospects, effective personalization efforts can nevertheless deliver right-time, right-message communications to make them aware of services that the bank offers. In the following section, we describe how personalization can help a firm achieve two broad marketing objectives:

- demand generation: building awareness, inducing trial or inquiry, and motivating the first purchase, and
- building the long-term value of customers: increasing their loyalty and retention, resulting in a larger share of wallet, or share of mind.

ACQUIRING NEW CUSTOMERS

According to the CMO Council's 2008 study, 47% of personalized communication is used to acquire new customers and 53% is used for retention and building loyalty.[100] Personalization requires data, and firms can get data about prospective customers in one of two ways: by getting customers to identify themselves, or by buying or renting lists of prospects who are in the target market.

First, let's see how firms find customers who are searching for their products. Today, customers can use an Internet search engine such as Google. Just type in the keyword, and up come advertising-supported sites on the right side of the Web page. The user clicks on a link and identifies himself or herself through a registration form or other request for information. This electronic method is simply the latest manifestation of how customers use "the marketplace." Traditionally, customers flocked to the physical location of stores, flea markets, local fruit and vegetable markets, boat shows at the mall, etc., in order to meet a specific need, or for the pure pleasure of shopping.

Business-to-business marketers have their own versions of this marketplace in trade shows.

To attract potential customers to shopping opportunities, consumer products companies still use mass media outlets to distribute advertising messages to a broadly defined target market. Interested customers are self-identified when they reply by dialing an 800 number, sending back a reply card, or visiting a Web site. An example of this approach was described in the *Inside 1-to-1 e-newsletter* published by the Peppers and Rogers Group.[101] Business trade magazines carried an advertising campaign for Jaguar, announcing the new XJ8L model. The print ads included a tear-out card asking interested customers to share information about themselves. Along with the traditional name and address, Jaguar also asked potential customers to identify at what stage in the purchase process they were, and what factors they considered important in determining the next car model they would buy. The campaign produced an immediate 40% jump in lead generation, and allowed Jaguar to understand its customers better. Jaguar advertising professionals went to great lengths in the wording of the card to capture the specific needs of each customer. For example, if a prospect only wanted to be mailed a brochure, a street address was requested but not an email address. In Jaguar's plan, mass media was initially used to find likely prospects, and it was likely that customized marketing communications would be sent next to move the prospects along their various paths toward purchasing a vehicle.

The second way to obtain information on prospective customers is to buy or rent a list from a list vendor. Direct marketers have used these resources for years to help create their own personalized campaigns through direct mail advertising and catalogs. Identifying, organizing, and distributing lists of prospective customers who share some common life stage or consumer history is the primary function of the list management industry. According to the DMA *2007 Statistical Fact Book*, just over 60% of the more than 200 direct marketing firms surveyed reported renting lists of prospective customers.[102] These prospect files are compiled by list vendors that group people or businesses based on specific demographic, lifestyle, or purchasing data. Prices to rent these lists vary considerably; from $273 per thousand names for B2B email contacts to $82 per thousand for a list of charitable donors.[103] Prospective customers on these lists are ones that have a predicted interest in a product as a result of past purchase behavior (e.g., the recent purchase of a home) or some correlated demographic descriptor (e.g., a high income and a newspaper subscription).

The nature of the data in a list determines the extent of the personalized message. In order to make customization relevant, firms need specific information about individual prospects. At the most basic level, the geographic location of a prospect can be used to create a customized version of a marketing message, which is the simplest form of personalization. An example of a versioned printed marketing program is presented in the following case study on Ace Hardware.

ACE HARDWARE INSERT TEST PASSES WITH FLYING COLORS

by Melissa Campanelli
DM News, Nov. 15, 2004, Volume 26, No. 43
Reprinted with permission of DM News, © Courtenay Communications

Thanks to a test insert program targeting a new consumer segment in May, Ace Hardware's Chicago-area stores saw paint and paint supply sales climb 24 percent compared with the same period in 2003.

The 4,800-store wholesale cooperative generally relies heavily on inserts because "they are cost-effective," said Colleen Donahugh, print, production, and coop manager at Ace Hardware, Oak Park, IL. "Inserts have a very broad reach at a very low price per impression."

Ace decided earlier this year that it needed help testing the effectiveness of its ad insert program and also wanted to test new customer segments. After consulting with Vertis, it decided on test targeting its newspaper inserts using psychographic data. Vertis, Baltimore, prints more than 40 percent of the 575 million inserts Ace distributes annually.

This was unusual because normally with insert advertising, "demographics and psychographics are not involved," Donohugh said. The ZIP codes surrounding a store usually form the basis of an insert program.

To find the appropriate new customer to target, Vertis used the research from its annual customer focus study, which tracks customer behavior across industry segments such as home improvement, furniture, grocery, sporting goods, and home electronics as well as media that includes ad inserts, direct marketing, and the Internet. Vertis then analyzed those groups in its RISC AmeriScan, which uses a personality segmentation system to analyze the socio-dynamic profiles of respondents. Vertis found an untapped customer segment that would be attracted to retail destinations offering home decorating solutions. The market was open, "slightly younger and more affluent than Ace's core customer, who tends to be middle class, middle income and slightly older," Donohugh said.

> The insert test targeted both groups: 40 percent of the roughly 2.5 million inserts went to Ace's core customers, while 60 percent went to the new target.
>
> Vertis then looked at newspaper zones and chose those with the highest potential for the core customers and for the new targets. Ace focused on the Chicago market, its largest U.S. market, with 155 retail stores. Newspapers used were the Chicago Tribune and the Chicago Sun-Times, along with their networks of suburban papers.
>
> Ace used different inserst for each segment. Both aimed to spur paint sales. They were designed to get customers into retail stores May 14-17. Identical products were highlighted, and both were 8-page inserts, but each had a different creative execution.
>
> "The core customer responds to price, product, and value, so their cover was a screaming deal of 'Buy 2, Get 1 Free!' which really emphasized the free offer," Donahugh said. "The other group, however, responds more to an idea that 'these are the types of things that I can do,' and had a woman on the front cover painting a wall."
>
> The color palette in each insert differed for each target group, she said.
>
> "The core customer responds to stronger colors while the new target responds to softer, more trendy colors, so those are the color schemes we chose for each group," she said.
>
> Donahugh attributed the bulk of the results to the new target group.
>
> "I think we connected with a target audience that generally looks to Ace as a place where they would shop for their paint and supplies," she said.
>
> "I think they think of us as a hardware store, and not necessarily when they are thinking of home décor."

Another example, reported in the Printing Industry Center's monograph *Investing in Digital Color... The Bottom Line*, was developed by Lexinet for Coldwell Banker.[104] The objective of this program was to generate additional real estate listings in the neighborhoods served by particular agents. A postcard was sent to neighbors after the listing of a home (see Figure 4.2). Local agents did not have the expertise to design, print, and find mailing lists of households in their regions, but relied on Lexinet to design a template where text and pictures specific to one agent's listing could be uploaded via the Internet to create a customized postcard mailing. Lexinet managed the printing process and mailing. Sometimes agents used their own lists of residents, but in many cases Lexinet provided rented mailing lists selected by ZIP code.

Personalization Strategies for Customer Development

The postcard could be customized by each agent with contact information and a photograph. The goal of the campaign was to build awareness of real estate agents within specific geographic areas so that when homeowners were ready to sell, they would remember these agents and call on them. In a way, the postcard became a pseudo-personal recommendation by a neighbor: "Use this agent, as we are doing, and sell your home fast and for what it's worth." Though the agent only knew the names and addresses of those who received the mailing, the information might have been perceived as highly relevant to the recipients because it referred to their own neighborhoods.

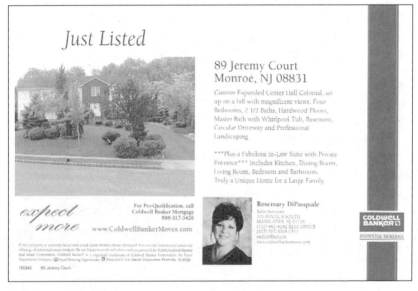

Figure 4.2 *Coldwell Banker customized direct mail postcard*

This type of personalization does not require a deep knowledge of the customer, and often uses simply the receiver's name and address. When personalized advertising is not perceived as valuable, it is often referred to as junk mail (or spam, in the case of email). In fact, many adults register on do-not-mail lists and buy spam-blocking software to avoid it. Consistent with this negative perception, the response rates to these communications are very low. When only a small amount of information is known about each prospective customer, the offer, message or promotion is not usually highly relevant to those who receive it.

In sum, a personalized marketing communications program for prospective customers should begin with an information acquisition strategy. A simple message sent to prospective customers (identified by their past history, geographic location, or life cycle stage) that is delivered by mass media, direct mail, email, or Internet search, and that elicits a reply, is a necessary first step in finding new customers who are ready to buy. These data are used to build a customer database that a marketer can leverage to create more sophisticated personalized offers in order to move the prospect towards purchase.

RETAINING AND BUILDING CUSTOMER VALUE

Once a customer has made an initial trial purchase, the firm must now deliver on its promised offer. Service is the starting point for building connections that will, in theory, increase the likelihood that customers will stay with the company. The first service encounter is the buying transaction itself.

Superior Customer Service and Transactions

A focus on customer service was the hallmark of the quality era of management thinking that grew in popularity in the 1980s and still persists today in the guise of Six-Sigma programs.[105] Customers will often mention a history of successful transactions when describing the characteristics of businesses they like to patronize.[106] Transactional exchange involves the transfer of something tangible or intangible between parties.[107] For intangible services such as banking, insurance, brokerage, or telephony, a printed record of the transaction between firm and customer is the tangible record of the delivery of the service.

The primary medium to deliver transaction information has been and remains the printed statement. According to the yearly U.S. Postal System's *Household Diary Study*, U.S. households received 100 billion pieces of mail in 2007.[108] That's a lot of printing! Included in this figure were over 18 billion bills and over 7 billion financial statements. However, online billing and payment are growing, due to the fact that these service are convenient for customers.

Once seen as a simple record of transactions from the seller's perspective, monthly statements can also deliver information about additional services for sale. For example, a telecommunications firm can detect how many in-state toll calls an individual customer makes, and may suggest that the customer switch to a new plan that features free local calls. (However, the monthly basic service fee may be higher. Read the fine print!) Monitoring the balances of its customers allows a bank to suggest additional financial products or services that fit the financial status of each customer. For firms with a frequent buyer program, personalized communications based on an individual's transactions will trigger notification of sales that shoppers may find hard to resist. Once the transaction interchange delivers satisfaction to the customer, building the commitment of the customer to the firm is the next step.

Building Lifetime Value by Developing Relationships with Customers

Fueled by the rhetoric of the 1-to-1 gurus, many marketers have been encouraged to build relationships with their current customers in order to improve profitability. Given the high cost of acquisition, more profit can usually be extracted by increasing customer retention rates (often used as a surrogate indicator of loyalty). Though some have questioned the link between profitability and loyalty,[109] others have described improved profits as a result of building loyalty through relationship marketing.[110] The basic theoretical constructs that are the foundation of relationship marketing theory can be found in this chapter's Appendix A.

The latest development in relationship marketing programs is called *dialogue marketing*. According to Kalyanam and Zweben (2005):

> Dialogue marketing is, to date, the highest rung on the evolutionary ladder from database marketing to relationship marketing, to one-to-one marketing. Its principle advantage over those older approaches is that it is completely interactive, exploits many communication channels, and is "relationship aware": that is, it continuously tracks every nuance of the customer's interaction with the business.[111]

What distinguished dialogue marketing from its earlier counterparts is its "attention to the temporal dimension of customer relationship...Dialogue-marketing systems are very sensitive to the interval between purchases, movement along loyalty curves, and increasing and decreasing frequency of physical and online visits."[112] For example, certain actions will trigger different communications tactics: one for first-time buyers, another for buyers who have purchased only once within a pre-defined time frame, and a third for those who have visited the online help desk with a question. These triggers are different for each type of business and come from a deep knowledge of the customer, the result of data-mining and predictive analytics. These analyses anticipate when a customer might become more valuable (e.g., increasing the dollar value of orders) or might be in danger of attrition (e.g., not making a purchase within a six-month time frame). The four types of dialogues described by Kalyanam and Zweben (the initial purchase interaction and three additional levels of interaction) are presented in Table 4.1.

Table 4.1 *Hierarchy of dialogues*[113]

FOUNDATION DIALOGUES:
1. Acquisition
2. Service follow-ups
3. Win-backs
LEVEL 1
1. Event notifications
2. Repurchase reminders
3. Inventory and price alerts
4. Overstocks
LEVEL 2
1. Defection interventions
2. Life Cycle progressions
3. Category Recency, Frequency, Monetary Value (RFM) transitions
4. Brand RFM transitions
LEVEL 3
For the future where on-site interactions will be easily achieved when customers are identified through scanning (voluntary keying of personal ID numbers) or RFID technologies (passive scanning of loyalty cards at point of purchase).

A successful example of Level 2 dialogue marketing is found in a 2003 article in the *Harvard Business Review* about Harrah's casinos.[114] Harrah's transformed the frequent user program it developed in 1997, "Total Gold," to one that was more personalized. Total Gold had been designed to provide regular customers with incentives to visit Harrah's casinos throughout the country. Slot machine players inserted their Total Gold cards into the machines and earned credits as they played, which rewarded them with free hotel rooms, dinners, show tickets, and gift certificates. The program was very similar to what other casinos were offering, however, and so did not build customer loyalty. As will happen with any undifferentiated program, the customers simply took their free benefits (in this case, rooms and dinners) and then moved on (in this case, to other casinos for more gambling). One benefit of Total Gold was that it provided a reliable stream of customer data. And so, at Harrah's headquarters, the data mining began.

In order to create a new incentive program, Harrah's computed the LTV (lifetime value) of each customer. The results were surprising—the best customers were not high rollers, but middle-aged and senior adults who had time on their hands, had comfortable incomes, and enjoyed playing the slots. Harrah's also found that satisfied customers had increased their gambling by 24% per year, while those who were unhappy decreased their yearly gambling by 10%.

Harrah's developed a number of incentive programs based on this knowledge. First, a series of direct mail communications was sent out, inviting lapsed customers to return. For example, if a customer who typically spent $1,000 per month hadn't gambled at Harrah's in a specific time frame, that customer would receive an invitation to a special event. If the same customer had lost money on his or her last visit, the event might be VERY special. Second, Harrah's implemented a three-tiered "customer care" program: Gold, Platinum, and Diamond cards were issued, based on customers' predicted LTV. Platinum and Diamond cardholders received higher levels of service, because Harrah's had realized what the best customers wanted. The article explains,

> They didn't want to wait in line to park their cars, or eat in restaurants, or check in at the front desk. So we decided to make a point of routing our customers into three different lines. People who weren't card-carrying Harrah's members and Gold customers stood in lines at the reception desk or the restaurant. Platinum customers would stand in still shorter lines, and Diamond cardholders would rarely ever have to stand in line. This created a visible differentiation in customer service.[115]

Gold customers could see that others were getting better service and so were motivated to earn higher points to move up to the Diamond service level.

The above example shows the power of personalization for businesses that directly interact with their customers. Firms that work through dealers or other value-added resellers can also take advantage of personalized marketing communications with the help of their geographically dispersed resellers. For example, at a local car dealership, the customization of a brochure could reflect local conditions (e.g., the season or time of year, local landmarks appearing in photographs, or offers that reference other local dealers or a specific customer base) and at the same time maintain a nationally consistent corporate look and feel, a requirement of international branding efforts. A manufacturer could also help dealers communicate with their customers through co-op advertising in local newspapers or television, point-of-sale displays or posters, or mailings targeted by ZIP code. With the correct infrastructure in place, a dealer could use Internet-distributed corporate display templates that would allow space for him or her to customize content as needed, and receive the resulting full-color, large-format posters within 24 hours of submission from a local print provider who has a contract with the corporate office.

In sum, relationship marketing is a viable strategy for firms that have sufficient information about individual customers. Based on gathered insights of how a customer uses products and services over time, firms can use timely personalized communications to build conversations with key customers. This will in turn lead to higher customer lifetime value (LTV) by keeping more customers, increasing the "share of wallet" of those customers, and decreasing the cost of doing business with them.

While success is built on the foundation of customer knowledge, the use of these tactics without sufficient insight leads to "failed relationship" marketing programs. As noted by Fournier, Dobscha, and Mick (1998), "The very things that marketers are doing to build relationships with customers are often the things that are destroying them."[116] When businesses have just a little information about a customer, such as a single previous purchase or a known geographic location, sending tons of email or junk mail is the primary tactic of the relationship-building program. As some marketers claim, over-communicating with customers is akin to stalking, and is received just as negatively. The next section describes how to use four types of personalization in the right ways to achieve marketing goals.

PERSONALIZATION TACTICS FOR CUSTOMER ACQUISITION AND RETENTION

As noted above, demand generation or acquiring a new customer involves building awareness, inducing trial or inquiry, and motivating a prospective customer's first purchase. This is often accomplished via the use of mass communications to solicit leads or responses by interested parties. Since not much is known about a prospect, simple forms of personalization are required. Retaining customers, on the other hand, can be facilitated by personalized communications that are sent in response to customer behavior as the firm tries to move the customer towards higher levels of commitment to its brand. Since more is known about an existing customer, more complex messages can be integrated into the personalized communications to make offers more relevant and to help build a dialogue.

Four categories of personalized communication tactics are presented below, from the lowest level of complexity (versioning) to the highest (fully-customized communications).

1. **Versioning:** These are often printed documents, such as catalogs or shopping inserts, distributed to groupings of individuals defined by a specific trait or characteristic. The document is not designed for a specific individual, but instead the content within is determined by that person's group membership (such as geography, gender, or language). An example of this is Ace Hardware's customized advertising inserts.
2. **Mail merge:** This type of personalization is named for early direct mail advertising, where the messages within letters were unique but that uniqueness may only have reflected the name and address of the recipient ("John, you may already be a winner"). If more is known about the recipient, such as the details of a recent purchase, more relevant content can be included in the message. Email delivers this kind of "behavioral targeting," frequently based on an individual's online search behavior.
3. **Transaction/Transpromotional:** Business documents such as bills, statements, orders, receipts, delivery notices, invoices, or shipping slips are used to make a record of a transaction. The personal information in documents from health care providers and financial services firms needs the security of accurate creation and delivery of the document to the intended recipient.
4. **Fully-customized communications:** This sophisticated publishing technique can vary every element on documents such as brochures, postcards, newsletters, or personalized Web sites or landing pages. Fully-customized communication enables organizations to create unique content for each recipient, including the ability to use different colors and graphics to make these documents stand out.

Table 4.2 links these four personalization levels with the marketing objectives of acquisition and retention described earlier in this chapter.

In *Data-Driven Print*, we included a personalization level called "Internet on-demand." In this scenario, document templates are distributed via the Internet to a number of document authors, who can then add personalized content. This might include the name and photograph of a local dealer placed in a document template that is managed by a corporate manufacturer. Internet on-demand is often used to develop dealer-specific documents such as sales collateral.

However, we no longer consider Internet on-demand a "level of personalization" since it falls into the aptly-named category of *Web-to-print* applications. Some, but not all, of these applications are also used as portals for

Table 4.2 *Levels of personalization and the typical marketing objectives*

LEVEL OF PERSONALIZATION	MARKETING OBJECTIVE	EXAMPLES
VERSIONING	ACQUIRE	One version of a gardening catalog is sent to households in the Northeast and another is sent to households in the Southwest.
MAIL MERGE	ACQUIRE	A new music store buys a mailing list of local households compiled from people who subscribe to a music magazine. All households on the list get a postcard announcing the opening of the store (event notification).
TRANSACTION/ TRANSPROMOTIONAL	RETAIN	A phone company sends an email suggesting that a customer move up to a new plan based on prior calling history.
FULLY CUSTOMIZED COMMUNICATIONS	RETAIN	A customized mailing is created based on past purchases and predictions of what new needs for similar or related items will arise. For automobile dealers, information about a new model of a leased car is sent to those whose lease is expiring soon (repurchase reminder). Or, a catalog retailer sends a notice of overstocks on the type of recreational equipment purchased by recipient.

variable data printing applications. These applications often act as a digital storefront for commercial printers and photo book printers. As implied by the name, Web-to-print applications (W2P) start with the Internet and end with printed material. Some commercial printers use W2P portals to give their business customers more options for ordering printed products. These can be as simple as uploading files to an FTP server, or as complex as an individualized inventory management system and variable data program for large, geographically distributed clients. In an article in the 2008 *NAPL Business Review*, author Howie Fenton explained that W2P can mean many things:

> It is important to recognize that "Web-to-print" is actually an umbrella term encompassing many different tools. It could include online estimating, a shopping cart for ordering, a file submission tool, a preflight tool, a proofing and annotation tool, and online design tools, or a strategy to merge variable data.[117]

A classification system for Web-enabled printing architectures was created by Adam Dewitz for his master's degree thesis at RIT. A summary of his work and a diagram of his typology are presented in Appendix 4B, later in this chapter. McKibben and Shaffer present another typology in their book *Web-to-Print Primer*. Their four categories of W2P systems are:[118]

- print procurement: e-commerce store fronts,
- marketing/brand management: franchise collateral ordering systems (like the Internet on-demand type of personalized printing described in *Data-Driven Print*),
- document management: mailing services, fulfillment and inventory management, and
- workflow automation: password-protected sites that allow clients to follow their jobs through the printer's production process.

A new resource to help printers find what system is best for them can be found at the Printing Industries of America (PIA)'s "Web2Print Test Drive Center," at http://www.w2ptestdrive.com/. Using the classification of systems by McKibben and Shaffer, the site lists over 40 vendors in the print procurement solutions category alone.

Whether you call it Web-enabled print or Web-to-print, the growth of online ordering of printing services cannot be ignored. For example, in the 2008 *Graphic Design USA* annual print design survey, 43% of GDUSA members (primarily graphic designers) reported ordering print online.[119] Whether these applications improve the printer's ability to sell personalized communication remains to be seen.

CONCLUSION

Creating and delivering personalized marketing materials is always related to a set of predetermined marketing objectives. When little or no information is known about a specific customer, an acquisition goal is appropriate, and the degree of customization will be low. When more information is known, building a relationship marketing program becomes possible. To meet this new goal, a dialogue plan must be articulated so that customized communications can be built for customer responses and non-responses. The next chapter provides an overview of the information technology that is required to produce personalized communication at the most complex level.

APPENDIX 4A
OVERVIEW OF RELATIONSHIP MARKETING

When designing a relationship marketing program, it is helpful to understand the history of customer relationship marketing (CRM) and the theories that describe the dynamics between buyers and sellers. The notion of relationship marketing migrated to the business and marketing psychology fields from organizational behavior and industrial marketing research, where interdependence between firms has been identified as the foundation of successful B2B alliances. Morgan and Hunt defined relationship marketing in 1994 as all marketing activities directed towards establishing, developing, and maintaining successful relational exchanges.[120] In defining this key construct, Morgan and Hunt drew from social and clinical psychology, namely, the social exchange theory and the marriage literature. In their model, commitment and trust were key mediating variables because they encourage exchange partners to preserve relationship investments, resist attractive short-term alternatives, and maintain the belief that partners will not act opportunistically.

In the B2B world, the relationship between firms is usually developed through face-to-face interactions over a long time period, fostering a sense of mutual interdependence.[121] However, there has been a healthy debate among researchers about whether relationship marketing strategy is appropriate for explaining B2C (business to consumer) exchanges. Sheth and Parvatiyar claim that the application of relationship marketing to the consumer is logical:

> When producers and consumers directly deal with each other, there is a greater potential for emotional bonding that transcends economic exchange. They can understand and appreciate each others' needs and constraints better, are more inclined to cooperate with one another, and thus, become more relationship oriented.[122]

Consumers' willingness to engage in relationships with businesses, they argue, is evidenced by their participation in loyalty programs. However, Sorce and Edwards have found that customers rarely described participation in a loyalty or frequent buyer program in their definitions of what a relationship with a commercial firm meant to them.[123] Furthermore, Pressey and Mathews found that relationship marketing concepts were associated with the patronage of personal service establishments such as hairdressers and recreation centers, where face-to-face interactions were the norm.[124]

APPENDIX

Transaction-driven marketing strategies are still important even now that CRM has become common. Coviello, Brodie, Danaher, and Johnston measured the frequency of relational exchange strategies used in business practices, and found that about one-third of firms use a relationship marketing strategy.[125] Firms that tend towards relationship strategies are often B2B service firms, where face-to-face interactions stimulate the growth of commitment, trust, and cooperation between buyer and seller. This is consistent with O'Malley and Tynan's hypothesis that the opportunity to develop relationships is "only feasible for high involvement products characterized by inelastic demand where regular interaction with consumers occurs."[126] For example, the vintage instrument enthusiast who will pay just about any price for a rare guitar would be a better candidate for relationship marketing than the parent who is shopping for an instrument for a child's possibly fleeting interest in playing a musical instrument.

APPENDIX 4B
WEB-ENABLED PRINT ARCHITECTURES[127]

by Adam Dewitz
Excerpted from his RIT Master's thesis *Web-Enabled Print Architectures*

New methods for specifying and producing printed products are emerging as print services providers seek to streamline order management, reduce costs, and improve efficiencies in print supply-chain management. These emerging print production models rely on system architectures that use Web applications to interface with highly automated print production workflows. The application of the Internet in print supply-chain management is not a new concept and has been previously investigated. However, little scholarly research has been published on Web-to-print or Web-enabled print production system architectures.

This summary is excerpted from a Printing Industry Center monograph, *Web-Enabled Print Architectures*, by Adam Dewitz, Master's degree graduate from RIT's School of Print Media. His research examined a number of print services providers utilizing Web-enabled print systems, and led to the development of an instrument for looking at Web-enabled print services providers. The instrument was then used to analyze a number of print services providers, providing insight into various approaches to developing the Web application processes of a Web-enabled printing system. The in-depth company analyses are not included in this summary, but may be found in the full monograph, available at **http://print.rit.edu/research**.

SYSTEM ANALYSIS INSTRUMENT

In order to compare the various Web-enabled applications deployed by print services providers, an instrument for analyzing the Web-based front-ends to a print production system was developed. The instrument was developed after a preliminary analysis of Web-enabled system architectures used within the printing industry, and it provides a list of system-independent attribute descriptions that can be used to describe the Web application of a Web-enabled print production system. The instrument is described in detail below and is summarized in Figure 4B.1.

Software Application Type

Software application type questions are used to determine where product specification and order management is taking place. Options included Web-based, desktop-based, and a hybrid Web- and desktop-based approach.

APPENDIX

- **Web-based application.** Web-based applications are software applications that run completely within the Web browser. They do not require any non-browser software applications or computer processes beyond the traditional Web-based client/server system methodology. Web-based applications require a persistent Internet connection during use.
- **Desktop-based application.** Desktop applications are compiled software applications that run natively on a local computer system and are independent of a Web browser. These applications provide all the functionality to specify products and manage product orders. A desktop application may have specific operating system requirements that limit platforms on which it can be deployed (i.e., a desktop application that is compatible with the MS Windows or Mac OS platform only). Desktop applications do not require a persistent Internet connection except when transferring data to the print services provider's production system.
- **Web- and desktop-based application.** A Web- and desktop-based application uses a mixed approach, relying on Web-based and desktop-based software applications to completely specify and order a product. Such a system may use a Web-based application to facilitate account and order management, and a desktop application to handle product design and specification. Depending on the portion of the application being used, a persistent Internet connection may be required.
- **Real-time WYSIWYG editor.** A real-time "What You See Is What You Get" (WYSIWYG) editor provides an accurate representation of the product during the product design and specification stage. The editor updates the product representation as the print buyer makes changes.
- **Preview-based WYSIWYG editor.** A preview-based "What You See Is What You Get" (WYSIWYG) editor provides an accurate representation of the product during the product design and specification stage. However, the editor updates the product representation only when the print buyer requests a product preview.

Knowledge and Skill Requirements

The knowledge and skill requirements questions help determine the complexity of the user interface by analyzing the skill sets that a user may need to have in order to successfully specify and order a printed product.

- **Knowledge of printing product specifications required.** This attribute aims to determine whether the print buyer using the system requires

any preexisting knowledge of printed product specifications. This may include knowledge of document design and layout principles, product limitations, printing process limitations, understanding of design or printing terminology, and other printing workflow-related skills or knowledge.
- **Requires special software.** This attribute determines whether all product specification and order facilitation is handled through the primary Web application, or if third-party applications are required to facilitate some part of the process. This may include software applications for handling design and layout such as Adobe InDesign, image and photo editing applications such as Adobe Photoshop, text editors and word processing applications, and file management applications such as an FTP client would use. Web browser plug-ins are also included in the special software definition.
- **End user.** The end user attribute provides insight into the intended audience of the Web application and into the complexity of the system. The end user can be a business (B2B model), a consumer (B2C model), or a combination of both.

Product Formats

The product format attribute describes what type of products the system is designed to produce and how the system handles product choice. Products can be catalog-based or completely customizable.

- **Catalog-based product formats.** Product formats are constrained to a catalog of product offerings. The products may or may not be template-based. Systems may or may not have systems logic in place to insure product quality. This may include restricting image or graphical elements from bleeding off a page or being placed in locations that will degrade product quality.
- **Customized product formats.** Product formats are not constrained. The printed products can be as unique as the job being ordered.

Digital Assets Input and Input Methods

The digital assets input attribute analyzes what types of digital assets are provided and how print buyers supply these digital assets to the Web application. This includes the file formats permitted by the system and the methods used to transfer the digital assets to the print services provider.

APPENDIX

- **Portable Document Format (PDF).** Does the system allow digital assets to be submitted using the Portable Document Format, including the standardized versions of the format: the PDF/X family and PDF/A?
- **Rich Text Format (RTF).** Does the system permit rich text content to be supplied in the Rich Text Format de facto standard word processing exchange format?
- **Microsoft Word Document (DOC).** Can Microsoft Word Document files be supplied?
- **Comma-separated Values (CSV).** Can data be supplied in the CSV format?
- **Cellular assets only (JPG, TIFF, ASCII Text).** Does the system only allow digital assets to be supplied in a cellular form such as images supplied on their own, or text elements supplied via ASCII text input via a form field?
- **HTTP asset submission.** Are digital assets transferred to the print services provider using the Hypertext Transfer Protocol (HTTP)?
- **FTP asset submission.** Are digital assets transferred to the print services provider using File Transfer Protocol (FTP)?
- **HTTP and/or FTP asset submission.** Can digital assets be transferred to the print services provider using the Hypertext Transfer Protocol (HTTP) and/or the File Transfer Protocol (FTP)?

Output Intent

The output intent attribute determines what the desired outcome of the printed product will be. The output intent ranges from that of a print buyer ordering a pre-specified product to that of a print buyer specifying a completely customized product.

- **Create a single printed product and purchase it.** Systems using this approach allow a print buyer to specify a printed product and purchase it through the Web application. The digital assets supplied by the print buyer may be archived by the system for future orders.
- **Create a document and merge variable data.** This system allows the print buyer to specify a printed product with variable data fields. The print buyer supplies a data list to populate the variable data fields. The digital assets supplied by the print buyer may be archived by the system for future orders.
- **Order managed documents from asset library.** This is a traditional print on-demand order management system for static printed prod-

ucts. A content owner uploads digital assets to a Web-based digital asset library. Printed products are produced as they are ordered. The print buyer making an order does not have to be the content owner supplying the original assets. Product customization is limited.

- **Use templatized documents to create product.** The print buyer uses pre-designed layouts and document templates to specify the printed product. The print buyer has limited options for personal customization.
- **Select print buyer-submitted digital assets to populate a templatized document.** The print buyer submits digital assets and uses pre-designed layouts and document templates to specify the final printed product. The digital assets supplied by the print buyer may be archived by the system for future orders.
- **Select print buyer-submitted digital assets to populate a customizable document.** The print buyer submits digital assets and uses customized documents to specify the final printed product. The digital assets supplied by the print buyer may be archived by the system for future orders.
- **Select stock assets to populate a templatized document.** The print buyer specifies a print product by selecting stock images or graphical elements from a digital asset library to populate a pre-designed layout or document template. The final product specified by the print buyer may be archived by the system for future orders.
- **Select stock digital assets to populate a customizable document.** The print buyer specifies a print product by selecting stock images or graphical elements from a digital asset library to populate a customized document. The final product specified by the print buyer may be archived by the system for future orders.
- **Submit print-ready documents and order product.** The print buyer submits print-ready documents that were created using design and product constraints specified by the print services provider. The product design may be completely customized or based on a template provided by the print services provider. The final product specified by the print buyer may be archived by the system for future orders.

Proofing

Proofing provides a simulation of the final printed product before it produced. Proofing is traditionally achieved by producing the printed product using either the same printing process or a process designed to mimic the production process. The proof provides a contractual agreement between the print

APPENDIX

buyer and the print services provider. Web-enabled workflows are designed to efficiently produce products in quantities of one or greater. Requiring a physical proof can defeat the purpose of using a Web-enabled print production system. Systems using virtual (soft) proofing methods replace the need for traditional physical proofs.

- **Virtual proofing.** The system utilizes virtual or soft proofing methods to simulate the final printed product. Product approvals are done virtually before final product production.
- **Hardcopy proofing.** The system requires hardcopy proofs of the product to be examined and approved before final product production.
- **No proofing services.** The system provides no formal proofing services or methods.

Business Transaction Complexity

The complexity of the business transaction can vary from simple payment by credit card to more traditional purchase orders or lines of credit provided by the print services provider.

- **Credit Card.** Payment for the print services can be done via a credit card.
- **Purchase Order.** Payment for the print services can be done via a purchase order.
- **Net Billing.** Payment for the print services can be done via lines of credit established between the print buyer and the print services provider.

Distribution

Distribution plays an important role in the print supply chain. If a printed product does not arrive at its destination, it has failed to provide value or communicate its message. Distribution can also define how a printed product is specified and produced. A system designed to produce products of a personal nature may differ from a system designed to make products available to the open market.

- **Direct distribution.** The print buyer specifies the product. It is produced and shipped directly to the shipping address supplied at the time of the order.
- **List distribution (push).** The system utilizes list distribution methods to produce the product order, and then ships the products to a list of

recipients specified by the print buyer at the time of the order.
- **Private distribution to order (pull).** The system uses a distribution method that permits only authorized buyers to order a print product. The content owner manages the authorization list.
- **Public distribution to order (pull).** The system uses a distribution method that permits open access to the printed product. Anyone interested in buying the product is permitted to do so.

Ancillary Services

The use of a Web-enabled print production system removes many of the activities that provide little or no value to the product specification, production, and fulfillment processes. However, use of these highly automated systems does not necessarily prohibit print services providers from providing ancillary services. There are opportunities to provide ancillary services that can be requested, purchased, and fulfilled directly through the main Web application or through a Web site or Web application integrated within the primary Web-based product specification system.

- **Storefront services.** Storefront services provide a Web-based marketplace where the content owners can sell their printed products. The print services provider furnishes the storefront and payment processing system.
- **Electronic product ordering.** A print services provider furnishes the infrastructure to facilitate electronic distribution in an electronic format, such as an eBook or some other digital file format. Digital Rights Management (DRM) services can also be provided for electronic distribution.
- **Digital asset management services.** Once a print buyer has submitted complete or cellular digital assets, the print services provider manages the digital assets.
- **Design services.** Graphic design services are available to enhance the print buyer's final product or to provide design concepts for products under development.
- **Publishing services.** The print services provider offers publishing services, such as editing, proofreading, International Standard Book Number (ISBN) distribution, and product marketing.
- **Distribution services.** The print services provider provides professional distribution services. This includes making the product available through wholesale channels and distributors.

APPENDIX

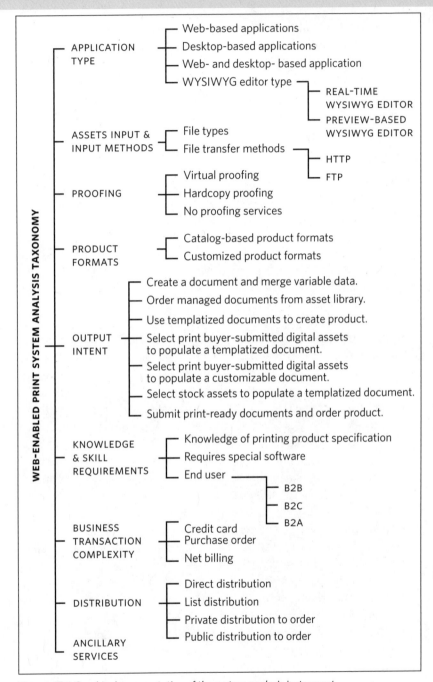

Figure 4B.1 *Graphical representation of the system analysis instrument*

APPENDIX

A graphical representation of the system analysis instrument is shown in Figure 4B.1.

SITE ANALYSIS

The goal of the site analysis was to determine whether the System Analysis Instrument provides the vocabulary to accurately describe the Web application used in a Web-enabled print workflow. Additionally, the site analysis provides an overview of seven different approaches to Web-enabled print. Although each Web-enabled print service provider analyzed took a unique approach to build its service offerings, there are some similarities between each system. See Table 4B.1.

Sites analyzed included the following (organized based on the product specification and distribution methods):

Order cataloged products with templatized customization using direct distribution:
- MagicPrints
- Blurb
- ShutterFly
- Moo

Order cataloged products with design/content customization using direct distribution:
- Lulu
- VistaPrint

Order cataloged products with templatized customization using list distribution (push):
- Cardstore.com

Application Type

A majority of the print services providers utilize completely Web-based applications to handle product specification and order management tasks. Blurb uses a desktop-application to handle product specification, but relies on a Web-based application to manage account and order management tasks. Rich Internet Application (RIA) technology has reached a point that many features and functionalities traditionally associated with desktop applications, such as drag-and-drop, can now be deployed within the Web browser. Many of the features found in Blurb's BookSmart application can be implemented

APPENDIX

Table 4B.1 *Tabulated results from site analysis*

SOFTWARE APPLICATION TYPE	COUNT	PROOFING	COUNT
WEB-BASED APPLICATION	6	VIRTUAL PROOFING	7
DESKTOP-BASED APPLICATION	0	HARDCOPY PROOFING	0
WEB- AND DESKTOP-BASED APPLICATION	1	NO PROOFING SERVICES	0
REAL-TIME WYSIWYG EDITOR	2		
PREVIEW-BASED WYSIWYG EDITOR	5	**BUSINESS**	
		CREDIT CARD	7
KNOWLEDGE AND SKILL REQUIREMENTS		PURCHASE ORDER	0
REQUIRES SPECIAL SOFTWARE	1	NET BILLING	0
PRODUCT FORMATS		**DISTRIBUTION**	
CATALOG-BASED PRODUCT FORMATS	7	DIRECT DISTRIBUTION	7
CUSTOMIZED PRODUCT FORMATS	0	LIST DISTRIBUTION (PUSH)	1
		PRIVATE DISTRIBUTION TO ORDER (PULL)	1
DIGITAL ASSETS INPUT AND INPUT METHODS		PUBLIC DISTRIBUTION TO ORDER (PULL)	1
FILE FORMATS	6		
CELLULAR ASSETS ONLY (JPG, TIFF, ASCII TEXT)	1	**ANCILLARY SERVICES**	
		STOREFRONT SERVICES	2
HTTP ASSET SUBMISSION	6	ELECTRONIC PRODUCT ORDERING	1
FTP ASSET SUBMISSION	0	DIGITAL ASSET MANAGEMENT SERVICES	7
HTTP AND FTP ASSET SUBMISSION	1	DESIGN SERVICES	1
		PUBLISHING SERVICES	1
		DISTRIBUTION SERVICES	

within the Web browser using standard Web technologies such as HTML, CSS, JavaScript, and XML. Embedded Web browser plug-ins can be utilized to address any shortcomings of the standard Web technologies. Completely Web-based applications provide a number of benefits over desktop applications. For example, bug fixes and new features can be silently released without having to update software on the user's computer.

Knowledge and Skill Requirements

All of the print services providers analyzed have built workflows that minimize the knowledge of printing product specifications required to successfully order some or all of their print services. Lulu provides both print services that can be used without any special skills and print services that are designed to meet the requirements of users with design and printing product specification knowledge. The analysis of all the systems shows that requiring no special software and using a completely Web-based approach almost eliminates all special knowledge and skill requirements needed to successfully use the system.

Product Formats

All of the print services providers analyzed constrain product offerings to a limited number of products and product formats. This approach is essential not only in building low-barrier-to-entry Web applications, but also in designing process-integrated production systems that can efficiently produce products.

Digital Assets Input and Input Methods

Digital asset formats are constrained to a select number of formats. PDF, JPEG, GIF, and PNG formats are utilized by a number of the systems. These are popular business and consumer formats, and a number of propriety and open source software libraries exist to manipulate these file types. While Hypertext Transfer Protocol (HTTP) can be a slower protocol for data transmission, it is being successfully used by all the service providers analyzed. Lulu provides an additional File Transfer Protocol (FTP) submission option for large files.

Output Intent

All the systems allow the print buyer to specify a printed product and purchase it through the Web. Using templatized documents to create product is the common approach. This enables the service provider to use a more streamlined order specification workflow. A highly constrained, template-

APPENDIX

driven workflow enables complete specification of the printed product in three or four steps. None of the service providers allow completely customized products to be specified. Building automated systems for these types of products is not commercially possible at this time.

Proofing

All of the print services providers use virtual (soft) proofing methods to replace the need for traditional physical proofs. None of the services offer hardcopy proofs. If the print buyer requires hardcopy proofs, a single copy of the product must be ordered.

Business Transaction Complexity

A business relationship can be established with all of the service providers over the Internet. This process is as simple as creating an account by filling out an HTML form. All the service providers require payment for the print services to be done via a credit card.

Distribution

The most common distribution method is to produce the product and ship it directly to the shipping address supplied at the time of the order. The book publishing services (Lulu and Blurb) also provide storefronts for content owners to sell their products. These two services use both private distribution-to-order methods that permit only authorized buyers to order a print product and public distribution-to-order methods that allow anyone interested in buying the product to do so.

Ancillary Services

A number of the print services providers offer digital asset management services to assist the print buyer in re-ordering products. However, the capabilities of these digital asset management services are limited. Most of the systems do not provide download access to the digital assets once submitted to the system. One popular service is providing Web storefronts that enable customers to sell their products. The print services providers furnishing these storefront services have further opportunities to generate revenue by placing a service fee on each product a customer sells.

NOTES

97. Payne and Frow, "A Strategic Framework for Customer Relationship Management."
98. Kotler, *Marketing Management*.
99. Payne and Frow, "A Strategic Framework for Customer Relationship Management."
100. CMO Council, "The Power of Personalization: The Impact and Influence of Individualized Content Delivery."
101. Peppers, "Jaguar Uses Mass Market to Learn About Individual Customers."
102. Direct Marketing Association, *Statistical Fact Book*.
103. Ibid.
104. - Pellow, Pletka and Banis, *Investing in Digital Color… The Bottom Line* (PICRM-2003-10).
105. Six Sigma was the name of the quality program developed at Motorola in the 1980s to reduce defects in production. See http://www.isixsigma.com/library/content/c020815a.asp for more on the history of Six Sigma programs.
106. Sorce and Edwards, "Defining Business-to-Consumer Relationships: The Consumer's Perspective."
107. Bagozzi, "Toward a Formal Theory of Marketing Exchange."
108. U.S. Postal Service, "The Household Diary Study: Mail Use and Attitudes in FY 2007."
109. Reinartz and Kumar, "The Mismanagement of Customer Loyalty."
110. Reichheld, *The Loyalty Effect: The Hidden Force behind Growth, Profits, and Lasting Value*.
111. Kalyanam and Zweben, "The Perfect Message at the Perfect Moment," p. 114.
112. Ibid, p. 115.
113. Ibid.
114. Loveman, "Diamonds in the Datamine."
115. - Ibid.
116. Fournier, Dobscha, and Mick, "Preventing the Premature Death of Relationship Marketing."
117. Fenton, "Profit Generator?" p. 49.
118. McKibben and Shaffer, *Web-to-Print Primer*.
119. "45th Annual Print Design Survey," 2008.
120. Morgan and Hunt, "The Commitment-Trust Theory of Relationship Marketing."

121. Iacobucci and Hibbard, "Toward an Encompassing Theory of Business Marketing Relationships (BMRs) and Interpersonal Commercial Relationships (ICRs): An Empirical Generalization."
122. Sheth and Parvatiyar, "The Evolution of Relationship Marketing."
123. Sorce and Edwards, "Defining Business-to-Consumer Relationships: The Consumer's Perspective."
124. Pressey and Mathews, "Barriers to Relationship Marketing in Consumer Retailing."
125. Coviello, Brodie, Danaher, and Johnston, "How Firms Relate to Their Markets: An Empirical Examination of Contemporary Marketing Practices."
126. O'Malley and Tynan, "Relationship Marketing in Consumer Markets: Rhetoric or Reality?"
127. Excerpted from Dewitz, *Web-Enabled Print Architectures*.

CHAPTER FIVE

DATABASE TECHNOLOGIES FOR PERSONALIZATION

THUS FAR WE HAVE IDENTIFIED two requirements for the use of personalized marketing communications. The first requirement is having a customer relationship marketing (CRM) strategy, built on the knowledge of individual customers or prospects. This will yield a theory of consumer behavior that will inform the marketing tactics designed to fit the way typical consumers search for, choose, and use a product or service. The theory of consumer behavior for each firm's customers provides a blueprint for the data required to segment and target customers for specific marketing programs and campaigns. The second requirement is the ability to harness the knowledge within customer databases to create customized marketing offers that can be distributed in print (as in personalized direct mail), via email, or through corporate Web sites. As shown in Figure 4.1 in the previous chapter, the data repository is at the heart of the CRM software systems that gained popularity with the approaching Y2K (Year 2000) fears of total computing collapse.

In this chapter, we examine the portion of the CRM system referred to in Figure 4.1 as "multichannel integration processes," or, as others have named it, "enterprise marketing management" or "marketing resource management" (MRM). The term *marketing resource management* denotes a broader construct, encompassing planning, project management, and content distribution.[128] Software vendors such as Unica or Aprimo are the market leaders for large businesses. Other software systems are smaller in scope and specialize in campaign management, variable data printing, fulfillment or Web-enabled print. Some examples of these vendors are Four51, Pageflex, XMPie, Exstream and InterlinkONE. This chapter presents the features found in these more focused marketing automation systems that emphasize the creation and production of personalized marketing messages utilizing information found in customer databases. Before we profile these systems, we begin with a general discussion of the strategic deployment of database technology for marketing purposes.

CUSTOMER DATABASE TECHNOLOGIES

Observers of business technology have seen the rise, fall, and resurrection of CRM system installations over the last decade. The hype that preceded the adoption of these systems promised a transformed business based on a single view of the customer across the entire enterprise. The reality of implementing these ambitious systems required extensive business process re-engineering that was not only expensive but also disruptive to employees, and in particular to sales professionals, who found that the systems interfered with the selling process.[129] In some cases, firms completely abandoned their implementation after many years and dollars had been spent trying to get the systems to work.[130]

The complexity of these implementations cannot be underestimated. A conceptual framework has been developed that delineates this complexity. Jayachandran, Sharma, Kaufman, and Raman (2005) identified five relational information processes that are enabled by CRM technology use. Relational information processes encompass "the specific routines that a firm uses to manage customer information to establish long-term relationships with customers."[131] These routines are described in Table 5.1.

Table 5.1 *Relational information processes*[132]

PROCESS	EXAMPLE
INFORMATION RECIPROCITY	"We enable our customers to have interactive communications with us"
INFORMATION CAPTURE	"We collect customer information on an ongoing basis"
INFORMATION INTEGRATION	"We integrate customer information from the various function that interact with customers (such as marketing, sales and customer service)
INFORMATION ACCESS	"In our organization, relevant employees find it easy to access required customer information"
INFORMATION USE	"We use customer information to develop customer profiles"

The authors posit that these information processes are enabled by a CRM system that, when built on a customer database, is able to deliver the following:[133]

- sales support(e.g., assigning leads to field sales personnel),
- marketing support (e.g., analyzing responses to marketing campaigns),
- service support (e.g., scheduling and tracking service delivery),
- analysis support (e.g., calculating customers' life time value, or LTV), and
- data integration and access support (e.g., integrating customer information from different contact points, such as mail, phone, and the Internet).

But getting closer to customers does not need to be this complicated. Rigby and Ledingham reviewed companies who have had recent success with CRM implementations.[134] They found these firms shared a common blueprint for success:

> They've all taken a pragmatic, disciplined approach to CRM, launching highly focused projects that are narrow in their scope and modest in their goals. Rather than use CRM to transform entire businesses, they've directed their investments towards solving clearly defined problems within their customer relationship cycle—the series of activities that runs from initial segmenting and targeting of customers all the way to wooing them back for more.[135]

Rigby and Ledingham claim that an enterprise-wide view of the customer creates a need for complete and perfect data about the customer, accessible in real-time to business decision makers. However, obtaining, updating, distributing, and securing that information is extremely expensive. Firms need to establish priorities about the kinds of information they need to maximize the customer's experience, based on factors that cause the greatest level of immediate dissatisfaction. In other words, what are the mission-critical data that are required to deliver promised benefits to customers? Taking an example presented by Rigby and Ledingham, a hotel front desk manager's top priority is to know real-time data on room availability; it is less important to know what the next customer in line for check-in thinks about the lobby's décor.

The reduction in scope of CRM activities to solving clearly-defined and high-priority problems is good news for the outsourced service providers of advertising, direct marketing, and printing. These vendors do not need to provide a full CRM solution (as shown in Figure 4.1) to firms who probably would not be able to get a satisfactory return on an enterprise-wide investment. However, they can help client firms automate selected parts of their marketing programs, which is a needed service. As mentioned in Chapter 3, the CMO Council found that over 40% of its survey respondents indicated that one of the challenges for integrating personalized communications into their marketing programs was the complexity of doing this work.[136]

At the heart of the complexity is the system architecture. The major components of an information technology system that can produce personalized marketing messages are presented in Figure 5.1.

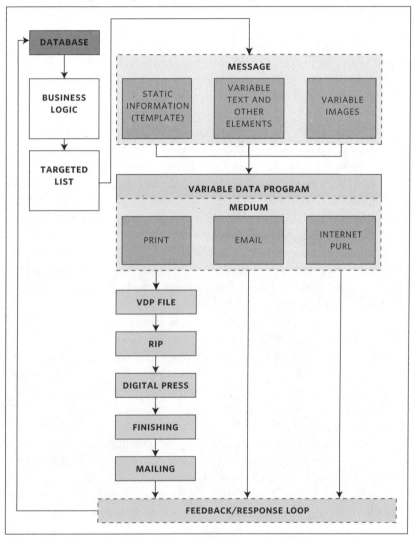

Figure 5.1 *A personalized campaign workflow*

It all starts with the database. With the wealth of information that can be contained in a relational database, a company can query a customer's order history and use data mining to discover insights that may lead to future marketing efforts, including what customized offers may be relevant to that customer. A list of customers who will receive a specific offer (the target market) is generated by utilizing conditional logic statements in data processing (e.g., selecting women members of an organization for a gift shop discount offer). A description of common logical operations is presented in Appendix 5A: Database Fundamentals.

Once the individuals are selected to receive an offer and the design of the promotional materials is completed, the content of the customized message is integrated into a variable data software program. As the market for personalization software has matured, more of what was once called "point solutions"—that is, software that has one primary functionality—have been broadened to include database management, campaign management, and online ordering (Web- enabled print). Appendix 5B highlights some of the variable data printing (VDP) software solutions available today.

To demonstrate how all of the pieces fit together, we present the following hypothetical case study to explore how a print services firm can develop a customer relationship management campaign for a business customer. This example uses the integrated solution provided by InterlinkONE that includes campaign management, Web-to-print, fulfillment, and sales management functionality, in addition to VDP software. For this hypothetical case, the print services provider is SourcePrint, and the client firm, WideWaters Gaming, is a company in upstate New York that offers horse racing and video game machines.

WIDEWATERS GAMING:
GROWING THE PLATINUM MEMBERS CLUB

WideWaters Gaming (our hypothetical business client) offers a wide range of entertainment services to the upstate New York region: thoroughbred racing daily from April through November, six-hundred video gaming machines (including poker) offered in a large gaming hall, and a tasty and elegant buffet. Visitors are encouraged to sign up for the Players club program that has three levels – silver, gold and platinum. The membership card acts as a debit/credit card required to play video games. This card also allows WideWaters to track the activities of players as they wager on horse racing, order food, and play video games. The Players club database includes these fields gathered during the registration process as presented in Table 5.2:

Table 5.2 *WideWaters Players' Club Database*

CUST. #	GENDER	LAST NAME	FIRST NAME	STREET ADDRESS	CITY	STATE	ZIP	YEAR OF BIRTH
102	M	Smith	Joe	123 Main St	Cary	NY	14511	1947
105	F	Macy	Anne	456 Oak St	Lima	NY	14555	1980

The transaction database records the activities of members as they enjoy the entertainment. The fields in this database might include the following as shown in Table 5.3:

Table 5.3 *WideWaters Visit Activity Database*

CUST. #	DATE OF VISIT	SVC1	SVC2	SVC3	SVC4	SVC5	WIN-NING	LOSSES
102	06/02/08	Race1 wager	Poker slot	Race2 wager			$0	$112
333	06/02/08	Penny slots					$23	$45

WideWaters has been served by SourcePrint (our hypothetical print services provider) for all of its commercial printing, from the racing forms to the member welcome kits. SourcePrint has recently purchased a production color digital press and is attempting to sell 1-to-1 marketing solutions to its customers. Jay Sullivan, VP of sales, is on friendly terms with Bill Leary, Director of Marketing and Customer Care at WideWaters. Bill also oversees other advertising tactics including weekly ads in the local daily newspaper and ads airing on cable television.

CASE STUDY

In preparation for his upcoming sales call with Bill, Jay has reviewed the marketing activities at WideWaters that are designed to build relationships with its customers with the goal of increasing their frequency of visits to the track, cross-selling (e.g., encouraging horse racing fans to try the video slots), and upselling (getting horse-racing fans to book viewing suites for parties and functions). The current member notification program is sending email blasts once a month alerting the members of special events, and placing posters in the track and game room to announce other services offered and the benefits of higher levels of membership.

The level of membership is dependent upon the number of visits to the track or game room. Everyone who registers starts at silver; gold is earned after five visits per year and/or an approved credit line of $2500. Platinum level is awarded after ten visits annually or a minimum of $5000 in credit. There are now 2500 members at the silver level, 3000 at gold, and 200 at platinum. The data managers at WideWaters have run analytics on the database and have learned that level of membership is highly correlated with annual amount of money expended at the track and game room. Bill has asked Jay to help him develop a marketing campaign to move more silver and gold members to platinum. After reading the case about Harrah's casino, he asked Bill is "Could there be tangible benefits in the platinum level that would be desired by the other members?" To get the answer, WideWaters conducted a number of focus group interviews with its members. They found that parking services were a very high priority for older members, while getting free merchandise in the track store was important to women. Based on this input, Bill revised the membership program reward levels to add "preferred parking" for gold (closer to the building) and free valet parking for platinum members. For shopping credits, each dollar spent on gaming provided a credit of 5 cents to spend on merchandise for silver members, and 10 cents for members at the gold level. These credits accrue on member cards automatically and can be read in kiosks located throughout the track, game room and restaurants.

Then Jay and Bill discussed a new campaign to alert members to these new benefits and to cross-sell and up-sell the additional services. They planned to continue the monthly emails but to also segment them by level of membership. They also planned to mail a personalized booklet to silver and gold members at the beginning and middle of each season to offer promotions tailored to their interests, reinforce the rewards program, and update them on how close they were to attaining a higher member level. After this strategy was developed, the next step was to design the materials. The graphic design of the new program was completed by WideWaters current

CASE STUDY

local ad agency. As this was underway, SourcePrint started working on the campaign mailing and printing logistics.

THE CAMPAIGN

First, SourcePrint asked for WideWaters to provide a database of its customers with the fields noted in Table 5.4.

Table 5.4 *WideWaters Database Provided to SourcePrint*

1.	Customer number
2.	Last name
3.	First name
4.	Street address
5.	Town
6.	State
7.	Zip
8.	Email address
9.	Year of birth
10.	Gender
11.	Member level
12.	Most recent visit
13.	Number of visits in last year
14.	Winnings (annual)
15.	Losses (annual)
16.	Number of horse race wagers
17.	Number of video games played
18.	Line of credit amount

With interlinkONE's ilinkMARKETING application, SourcePrint was able to create the details of the personalized booklet campaign in the system. Throughout the campaign, the system can track and report all activity. See Figure 5.2 for the personalized booklet campaign planning page.

First, the customer data is imported into the marketing database via the contact import feature (See "Add Source" in Figure 5.2). Because the system is integrated, each module will be able to access that data. For this application, the variable data print component in the system uses the data to create print-ready files for the personalized booklet. Other modules of the InterlinkOne system can use the email address field in the database to

CASE STUDY

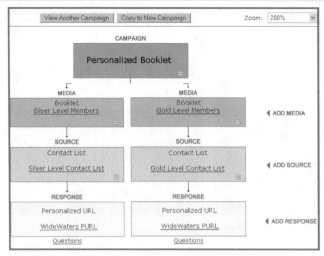

Figure 5.2 *The personalized booklet campaign planning page*

send follow-up emails to non-respondents. The booklet requires the following fields:

- first name
- member level
- number of visits in the previous year
- number of horse race wagers
- number of video games played

These fields determine the text, images, and colors for each member's booklet.
    Next, the graphic design team designs the booklet with programs such as Adobe Photoshop, Illustrator, and InDesign. In our example, the front cover of the booklet contains two variable fields of the customer's name and an introductory paragraph (See Figure 5.3). Once the booklet is designed, it can be uploaded into interlinkONE's Variable Data Print Engine where its variable fields are tagged for customization (See "Add Media" label in Figure 5.2). The VDP Engine merges the template with the customer data and produces a customer ready print file.

CASE STUDY

Figure 5.3 *First page of personalized booklet*

Figure 5.4 shows the inside of the booklet contains four variable fields:

- interest articles: From the information received from a customer's activities or membership survey, it can be determined what attractions will interest that customer. Articles can then be inserted that appeal to each individual.
- variable images: Images are then tied to interest articles. For example, if the article is about horse races, an image of a horse race will appear.
- membership status: The membership status box changes colors depending on what the next level is for that particular customer. A silver member who soon could be gold would have a gold box.
- personalized URL: At the personalized URLs, customers are asked for additional information to further target their interests. Here customers also find more information about the various benefits of higher levels of membership.

The back of the booklet contains a variable field that reminds customers to visit their Personalized URL.

The last step is designing the personalized landing page (PURL). Working with the graphic designer, SourcePrint creates landing pages with a similar look and feel to the booklet using interlinkONE's Landing Page Builder feature. After picking a template in the feature, SourcePrint can edit text, upload graphics, and add survey questions (See Figure 5.5 for a sample survey). When a member accesses the landing page, the system knows who the person is, and is able to greet him or her accordingly. The landing page is

CASE STUDY

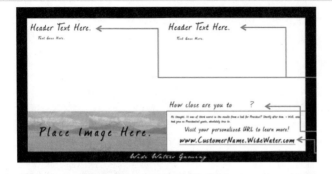

Figure 5.4 *Personalized interior pages to booklet*

Figure 5.5 *PURL survey*

CASE STUDY

connected to the imported customer database and is generated on-demand when the member visits the landing page, which is personalized with pictures, text, and the offer.

IMPLEMENTATION

Once all of these elements are uploaded, the campaign is ready to be implemented. SourcePrint prints and mails the booklets by the targeted date. Real time results are available through the "dashboard" feature as shown in Figure 5.6. WideWaters Gaming's marketing VP Bill Leary can log on to the system to determine how effective the mailing has been by accessing various graphs and charts. He can see who responded to the survey and what they said. Figure 5.6 shows that, of the 345 people sent the booklet, 92 (26.6%) opened the PURL and 46 (13%) filled out the survey form.

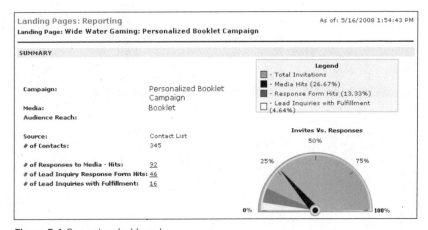

Figure 5.6 *Campaign dashboard*

When a member responds to the survey form, an email is created and sent to the WideWater sales team (See Figure 5.7) notifying the sales rep of the response. Thus, the WideWaters sales rep can take immediate action by reaching out to the member with any information that was requested.

Depending on how the member answers each question, different follow up contacts can occur to further the connection between WideWaters and its members. For example, if one member answers the question "How often do you think you'll be visiting us this year?" with "3-5 nights," a sales rep might inform the member of upcoming discounts on hotel accommodations. Or, if another member indicates that the parking was inconvenient, a personalized PDF can be sent explaining how there is free valet parking for platinum

CASE STUDY

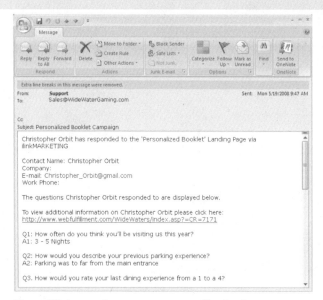

Figure 5.7 *Automatic survey response notification form*

members. If a member doesn't respond to the survey, then the system would generate an email notification in 3 weeks to re-introduce the personalized landing page feature. See Figure 5.8 for an example.

Figure 5.8 *Follow-up mailing to non-respondents*

CLOSING THE LOOP

Once the campaign is over, the InterlinkONE system can create reports of the response activities for each member in the database. For example, every day

Database Technologies for Personalization 97

CASE STUDY

at 9 am, InternlinkONE can send WideWaters an email that contains a link to an Excel file that lists recent leads. The WideWaters IT department can then update its database with the latest information on customer responses to the campaign.

CONCLUSION

This chapter provides a high-level overview of the database technologies needed to implement a successful personalized marketing campaign. The basic input necessary for personalization is data about individuals and their buying behavior. Once a marketing objective has been formulated, the personalization strategy should answer the question, "How can I begin or continue a dialogue with targeted customers or prospects?" For example, what behavior (or lack of behavior) by the customer should trigger a reminder postcard or email? Or, at what point should a customer receive the "gold card" loyalty benefit of free parking? Once specific triggering conditions have been articulated, customers can be identified within the database and selected, using conditional logic operations, to become the target group for a personalized marketing campaign that will include unique message content (text and images) for each individual. An integrated system will be able to create the campaign and send it by email, postal mail, or personalized Internet landing page. Once a particular customer's response (or lack thereof) is captured, the system will automatically update the customer database.

One important part of this process that we have not explored in depth is data mining—the engine that uncovers the insights about consumer behavior that reside in databases. Data mining in its most sophisticated form utilizes such complex statistical procedures as predictive modeling, cluster analysis, and artificial neural networks. These procedures are best suited for extremely large databases, typical of retail grocery stores. The automatic analyses that make up the larger data mining applications can produce insights about millions of customers buying thousands of products across multiple visits. The "best in class" campaign management and personalization programs automatically track customer responses and evaluate them. On the other end of the complexity scale, data mining can also deliver the ability to count the number of people who respond to a mailing or click on a particular URL. The use of simple descriptive statistics for assessing the outcomes of marketing programs will be explored in more detail in Chapter 8.

APPENDIX 5A:
DATABASE FUNDAMENTALS, BY MICHAEL PLETKA

The crucial starting point is the data. Often the data are provided by the client. Otherwise they are purchased or rented from list vendors who can provide a target population for a client's marketing needs. Experian, Dun & Bradstreet, and Hoover's are companies that sell or rent prospective client lists and provide demographic overlays to existing customer files.

The quality of data can make or break the success of a variable data print (VDP) application. In the 2008 CMO Council survey, 46% of respondents indicated that the lack of customer data and insight was a challenge to implementing their personalized marketing programs.[137] Print services providers who use client-supplied internal data need to be aware of the processes the client uses to ensure data quality. These processes include data cleansing, scrubbing, and updating. Quality assurance is also a factor when buying data. When purchasing or renting data from a list vendor, an important criterion is determining how recently that data was updated. Making sure that customer data in a database are correct and accurate can be a very time-consuming and resource-intensive process.

Another element of the quality of data is *degree of specificity*. For example, if a firm has customer names but no transaction data, it will not be able to predict when or what those customers might buy again. There are two criteria that define the usefulness of a database:

1. The database must contain all the data that are needed to execute the application successfully.
2. The database must be in a file format that the software can understand.

The nature of a firm's customer data will also determine what can be learned by *data mining*, the "iterative process of identifying previously unknown relationships and patterns in data to solve a business problem."[138] Not only must the data be accurate and specific, it must also be organized in a way that can be used in common applications. The database concepts that VDP users need to understand are explained below.

COMPONENTS OF A DATABASE

There are three main objects that comprise a typical database: the *data files*, *records*, and *fields*. A database can contain one or more data files. The data file

APPENDIX

is an entity that contains information about the objects in it. Objects inside a data file are called records. The attributes of a record are its fields. In other words, fields are the particulars of a record. If, as is pictured in Figure 5A.1, records are seen as the rows of information in a data file, then the fields are the columns.

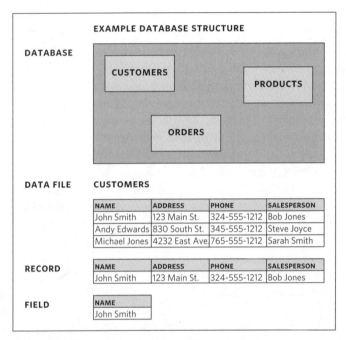

Figure 5A.1 *The basic structure of a relational database*

Figure 5A.1 depicts a database that could be used by retailers to track customers, products, and orders. The database would contain several data files, such as a file of products that the company has available for sale, a file of all of its current and past customers, and a file of the orders that those customers have placed. The data files would contain records with information about each individual customer, product, and order. In the records would be fields such as item numbers, item descriptions, order numbers, order totals, customer names and customer shipping addresses.

RELATIONAL DATABASES

If more than one data file exists, as is shown in Figure 5A.1, there is usually a relationship between some or all of the data files, in which case the database is called a *relational* database. A data file is one table or level of a relational database; these so-called *flat* data files are the basic building blocks of relational databases and are often simple text files or worksheets in a spreadsheet program.

ORDER

NAME	ADDRESS	PHONE	SALESPERSON	ORDERNUMBER	ORDERDATE	QUANTITY	PRODUCT	TOTALPRICE
John Smith	123 Main St.	324-555-1212	Bob Jones	12334	9/25/2003	3	Widgets	$100.87

Figure 5A.2 *A single record in a data file*

Figure 5A.2 shows an example of a record from a flat file database. While flat file databases are suitable for simple mailing or product lists, they are not the optimal solution for a company transaction database. Although the record in Figure 5A.2 is very simple, one can imagine the problems that may arise from a database with just this type of records in a single data file. For example, there may be multiple John Smiths in the database. If a customer made repeat purchases over time, it would not be optimal to continue to add more orders on a single record. Most customer databases overcome these complications by assigning each customer a unique number. Another complicating issue in a flat file database is ease of access to information. In the record shown in Figure 5A.2, the fields could easily reach 100 columns in width, making the database harder to navigate. The solution to all these quandaries is the relational database.

A relational database is actually a set of data files that are linked through the use of *primary keys*. Primary keys are special fields that can identify unique records in a data file. In the relational database in Figure 5A.3, the primary keys are "CustNumber," "ProdNumber," and "OrderNumber," acting as unique identifiers for each record in their respective tables. In the bridge table, "Orders," a primary key is assigned to each order, and then the appropriate customer and product keys are entered to link the information between the tables. Other information solely related to the ordering function, such as the quantity of products ordered, can also be included in the Orders file. The relational database structure is more efficient than a flat file database since it eliminates duplicate information and maximizes data integrity through the use of unique identifiers for each record.

Even with a relational database, issues with data integrity remain a concern. Data entry errors may create multiple "customers" in the database who are actually the same individual, making it difficult to determine which record contains the correct data. For example, John Smith lives at 123 Main

APPENDIX

CUSTOMERS

CUSTNUMBER	NAME	ADDRESS	PHONE	SALESPERSON
100	John Smith	123 Main St.	324-555-1212	Bob Jones
101	Andy Edwards	830 South St.	345-555-1212	Steve Joyce
102	Michael Jones	4232 East Ave.	765-555-1212	Sarah Smith

PRODUCTS

PRODNUMBER	PRODDESC	UNITPRICE
678	Widget	$99
890	Bookmark	$20
921	Notebook	$15

ORDERS

ORDERNUMBER	CUSTNUMBER	PRODNUMBER	QUANTITY	TOTALPRICE
1233	100	678	3	$297
2356	101	890	4	$80
5487	102	921	7	$105

Figure 5A.3 *Examples of data files in a relational database*

Street, his customer number is 100, and he has placed multiple orders with the company. On one occasion, an order representative mistakenly enters him as customer number 304. The impact of this duplicate information may be insignificant, such as John later receiving two pieces of the same correspondence in the mail. However, if the error affects billing information, a bigger problem will be created.

A process called *de-duping* can help eliminate such errors in a database. De-duping evaluates records, determining which ones are most likely duplicates that should be purged from the database.[139] Firms that purchase lists from vendors and integrate them into their internal data have to be aware that some customer records will overlap. De-duping will eliminate such duplication.

BUSINESS (CONDITIONAL) LOGIC

A fundamental assumption in creating personalized marketing messages is that different customers will receive different offers. For example, a firm may create one offer for customers who have not purchased anything in the past six months, and a different offer for customers who have made a purchase more recently. To match the offer to the customer, *conditional logic* is used.

Unfortunately there is no common language for programming *business logic* or *conditional expressions* across all variable information design programs. However, a basic understanding of computer programming is all that is needed to learn the functions. The following sections explain the most common programming functions encountered when designing variable data applications. Each example is written in Visual Basic code and then in Microsoft Excel. Visual Basic is one of the easiest plain-text programming languages to read and understand, and some variable data programs use codes similar to it. Microsoft Excel functions are an excellent way to test and model conditional logic without having any prior programming knowledge. Some other variable data programs create their business rules in syntax similar to this as well.

IF / THEN

The *IF statement* is the most common function used when creating business rules in a VDP application. Its basic operation is to test for a condition, and then do one of two actions, depending on whether the test turns out to be true or false. Many other functions can also be incorporated into an *IF/THEN* example, including some that are described in the next section. In the example below, the program is testing to see what value is in the "Gender" field in a database, in order to perform some action based on that value. If the person's gender is male, the program will output the text "Mr." for that record. If not (*ELSE*), the program will "know" that the person is a female and therefore will output "Ms."

<u>Visual Basic</u>
```
IF Gender = "Male"
      THEN "Mr."
      ELSE "Ms."
END IF
```

<u>Excel</u>
```
=IF(Gender="Male","Mr.","Ms.")
```

Nested IF

A *nested IF statement* is simply one or more IF statements included within an original IF statement. This is most commonly used when there are more than two possible outcomes for a given test. For example, let's say a person's favorite color (of three choices—red, yellow, or blue) is captured and coded in a data file record. We will first test to see if the person's favorite color is blue, and if it is, the program will output the "bluebox.tif" image for that record. If not, we have to determine whether the person's favorite color is red or yellow,

APPENDIX

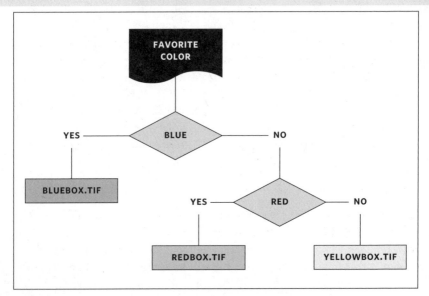

Figure 5A.4 *Example flowchart of a nested IF statement*

so we need to *nest* an IF statement within the original statement. This nested IF statement will test whether the person's favorite color is red, and if so, output the "redbox.tif" image. If the second test is also false, it is not blue or red and so must be yellow, in which case the system will output the "yellowbox.tif" image for that record. See Figure 5A.4 for a flowchart of this query.

<u>Visual Basic</u>
```
IF FavoriteColor = "blue"
      THEN "bluebox.tif"
      ELSEIF FavoriteColor = "red"
      THEN "redbox.tif"
      ELSE "yellowbox.tif"
END IF
```

<u>Excel</u>
```
=IF(FavoriteColor="blue","bluebox.tif",
IF(FavoriteColor="red","redbox.tif","yellowbox.tif"))
```

AND/OR Operators

An additional way to test for multiple conditions within an IF statement is to add *AND* and/or *OR* operators when there are two or more characteristics

APPENDIX

that will determine the outcome. In the following example, an AND operator is used within an IF statement to test whether the record is a male or female and under or over age 18. The outcome of these statements will determine the image that is embedded in the output. The program will then output the appropriate image as specified in the syntax (boy child versus adult male, for example). If the record doesn't return both tests as true, it will be considered a false value, and then a default image (e.g., group photo of men and women of all ages) would be included in the output.

Visual Basic
```
IF Gender = "Male" AND Age > 17
     THEN "male_adult.tif"
     ELSE IF Gender = "Male" AND Age < 18
          THEN "male_child.tif"
     ELSE "group_photo.tif"
END IF
```

Excel
```
=IF(AND(Gender="Male",Age>17),"male_adult.tif",IF
(AND(Gender="Male",Age<18), "male_child.tif","group_
photo.tif"))
```

OR operators work in the same way that AND operators do, except only one of the two conditions in the test must be true for the statement to return a true value. In our next example, if a person's age is coded in the record as either less than 25 years old or greater than 50 years old, the person will receive the "yellowcoupon.tif" image. A person who is between 25 and 50 years old, on the other hand, will receive the "greencoupon.tif" image.

Visual Basic
```
IF Age < 25 OR Age > 50
     THEN "yellowcoupon.tif"
     ELSE "greencoupon.tif"
END IF
```

Excel
```
=IF(OR(Age<25,Age>50),"yellowcoupon.tif",
"greencoupon.tif")
```

APPENDIX

Output Formatting / Transformations

Often the format of data in a field may not be the appropriate format to use readily in a VDP application. For example, most dates and times in a database are stored in the mm/dd/yyyy HH:MM format. This would be acceptable for an application such as the time and date an invoice was created, but not if a firm wanted to indicate a customer's birthday. This is where a *format* function is used to convert the display of data in a field into the most useful form. In the example below, this format function takes the "date" field from the database and converts it into a more readable form. If the date in the database is 10/14/2009 14:23, the system will change it and output just "October 14." Each variable data program has its own unique syntax, but the concept is generally the same across applications.

Another formatting operation is called *concatenation*. When displaying information in a variable data application, it is often useful to combine, or *concatenate*, unique fields into one text string. In the example below, we are combining the FirstName, LastName, and Age fields of a record into a sentence. If the current record is John Smith and he is 34 years old, the output would read "John Smith is 34." This is a relatively easy statement to create, but the process becomes tedious if the proper spaces are not included between fields or after constant strings included in the concatenated string.

<u>Visual Basic</u>
```
newString = FirstName & " " & LastName & " " & "is " & Age & "."
```

<u>Excel</u>
```
=CONCATENATE(FirstName," ",LastName," ","is ",Age,".")
```

These types of conditional logic statements identify the action to be taken to produce a customized offer. Depending on the complexity of the campaign, the variable elements of the communication may range from simply a name to more elaborate customization where unique graphics and text are selected. Variable information elements are combined with static elements to create the final piece. Combining the static and variable elements is further discussed below.

Static and Variable Elements

Once the database and business logic are prepared, the next step in creating a variable data application is the acquisition of all of the static and variable el-

APPENDIX

ements that will be used in the document. The static information, or the *shell* of the document, is often created in a page layout program such as QuarkXPress or Adobe InDesign. These blocks of static text and images will remain the same for each record. Examples of static elements could be a company logo, a disclaimer, or the terms and conditions for an offer.

The rest of the items in the document are variable elements, those that change depending on the record and business logic identified. Variable images can come from a number of sources, such as stock photo libraries, existing images from a company's digital asset management system, or new original images created by the company. Variable text elements can be developed by the company's marketing department, and may include headlines, subheadlines, and body copy elements. Some variable data programs also allow the creation of data-driven charts and tables. The creator must then ensure that the correct data is available in the database to generate these graphics.

Variable text files are useful when a firm has a large amount of text that will vary with each offer. For example, the specifications related to feature products will vary when each recipient receives information about a different product. If one record needs information about "Item A" in its VDP communication, a text block with those specifications and a description of "Item A" would be used. Once all of these elements are identified and collected, they can then be incorporated into the VDP application in a variable data program.

APPENDIX 5B: OVERVIEW OF FOUR VARIABLE DATA PRINTING SOFTWARE PRODUCTS

by Tracy Destino; Excerpted from her RIT Master's project *Variable Data Printing: An Exploration of Four Software Applications*

Once the static and variable text and graphic assets have been created for the campaign—as well as the business logic to dictate how these assets will be used with the accompanying database—the variable data software program brings all of these elements together to create a variable data document. There are many software products on the market today. Table 5B.1 below lists 20 different vendors available as of summer 2008; the list is not exhaustive.

Table 5B.1 *A list of variable data software programs*

VARIABLE DATA SOFTWARE PROGRAMS	NAME OF COMPANY (IF DIFFERENT)	COMMENTS
Agilis	Saepio	
DesignMerge Pro	DesignMerge	
DirectSmile		
EFI		
EskoArtwork		
Extream	HP	
Four51		Integrates with SAP
Fuse	L2 Soft	
Fusion-Pro Web		Printable integrated with Kodak EMS software
Group1		
Indigo Yours Truly	HP	
InterlinkONE		
Kodak Darwin		
LS soft		
Pageflex	Division of Bitstream	
Patternstream	Finite Matters	
Printshop Mail	Objectif Lune	
GMC		
XMPie	Xerox	
xPression3	EMC Document Sciences	
xPublisha	Sansui Software	

APPENDIX

These programs vary by complexity. Some variable data programs are designed for specific application types, such as creating billing statements or other transaction documents, or creating marketing collateral. The main features and distinguishing characteristics of four popular programs—Pageflex, Darwin, Printshop Mail, and XMPie Udirect—are described here, based on the criteria defined in Table 5B.2. Summary profiles for each program follow, in Tables 5B.3 to 5B.6.

Table 5B.2 *VDP application matrix: Software features and definitions*

SOFTWARE FEATURE	DEFINITION
ASSET FEATURES:	Any special features involving asset (image, text, graphic) placement and maintenance.
COLOR MANAGEMENT:	Does the software have any color management capabilities?
COST OF PROGRAM:	How much does the initial software purchase cost?
DATABASE CLEANING:	Does the software offer any technology to clean or maintain database?
DYNAMIC ELEMENTS:	What elements allow variability?
HARDWARE REQUIREMENTS (MAC / WINDOWS):	What hardware requirements are needed to load software?
LANGUAGE SUPPORT:	What languages/translations does the software support?
MAIL FORMATS – DESIGN OPTIONS:	What layouts, templates, formats, variable output materials are offered?
MERGE CAPABILITIES (Y/N):	Can software merge with existing database?
OUTPUT:	What options are available for printed output? Examples: postcards, brochures, letters, statements, etc.
PLUG-IN INTERFACE:	Are there any additional plug-in options available? What is their functionality?
POSSIBLE COST ADDITIONS:	Are there any additional software add-ons available at an extra cost?
PREFLIGHT:	Does the software offer any preflight technology?
PROGRAM SUPPORTED – DESIGN SOFTWARE:	What design software is the application compatible with?
PROOFING CAPABILITY:	Does the software offer any document proofing?
RIP SUPPORT:	Does the software include RIP support?

APPENDIX

SOFTWARE REQUIREMENTS:	What software requirements are needed to load the software?
SPECIAL FEATURES:	Are there any special features included in this software?
VARIABILITY LIMIT:	Is there a limit to the number of variable data elements per document or limit to size of database?
VDP LOGIC:	What is the method of assigning variable data elements to the design?
WEB OUTPUT - PRINTING/URL/EMAIL:	Does the software offer any cross-media or web output?

PAGEFLEX

Pageflex by Bitstream is a variable data printing program that works both as a stand-alone program and as a design plug-in. The customer/client can begin a design in either option; however, it has to be on a Windows-based computer platform.

Pageflex offers unlimited variability customization on all of its applications, including the ability to add previously created PDFs to the file. Common variable elements include graphics, text, and fonts. Other customization options include:

- color schemes,
- size and position of the variable elements,
- page numbers and size,
- appearance and attributes of various design elements,
- output and delivery methods, and
- both if/then logic and more sophisticated scripted database logic.

Documents are personalized through a link to a database. The database can be created in a variety of forms, ranging from a simple ".csv" format to an Access database. The program also accommodates more elaborate database programs such as Oracle and MySQL that allow for more complicated variable logic. This feature would be useful to a large company that has lots of information on its customers.

Pageflex also offers features intended to keep the flow of the document consistent. The program can distribute white space on each page to evenly place all items. It can adjust the size and/or shape of an image if the text flow on the document changes, eliminating or precluding any problems with text overflow (a feature called "autotext"). Additionally, the program can easily adjust the format for more variable elements to be added after the

APPENDIX

initial design has been determined.

Pageflex also offers additional products that can expand its functionality to include variable campaigns. This allows the creation of not only single documents, but also campaigns that deliver consistent and personal messages to customers via, for example, personalized Web pages and text messaging. These multi-channel campaigns are becoming a more frequently used and sought-after solution.

Pageflex offers a lot of flexibility. Although Bitstream does not offer online, downloadable software to test, it does offer a pre-purchase hands-on demonstration with a company representative. Bitstream's training center in Maryland caters to those who have purchased Pageflex.

Table 5B.3 *Pageflex analysis*

ASSET FEATURES	Auto Text / Image fitting
COLOR MANAGEMENT	Information not available
COST OF PROGRAM	Information not available
DATABASE CLEANING	Supports almost all database formats.
DYNAMIC ELEMENTS	Text, Images, Color, Pages, Layouts
HARDWARE REQUIREMENTS -MAC	Information not available
HARDWARE REQUIREMENTS - WIN	See Below
LANGUAGE SUPPORT	English, Text can be composed and output in more than 60 languages.
MAIL FORMATS- DESIGN OPTIONS	Database-driven personalization, Online Design and Editing, Form-Driven customization, Cross Media Campaigns
MERGE CAPABILITIES Y/N	Yes
OUTPUT	Standard Output Drivers: PostScript, Optimized PS, Adobe PDF, EPS
PLUG-IN INTERFACE	Various PageFlex add-ons
POSSIBLE COST ADDITIONS	Various PageFlex add-ons
PREFLIGHT	Yes - with job batching
PROGRAM SUPPORTED - DESIGN SOFTWARE	QuarkXPress, Adobe InDesign, PageFlex Studio
PROOFING CAPABILITY	Yes

Database Technologies for Personalization

APPENDIX

RIP SUPPORT	Yes
SOFTWARE REQUIREMENTS	MS Windows XP Professional, Windows Server 2003
SPECIAL FEATURES	Flexibility to change size and position of elements automatically.
VARIABILITY LIMIT Y/N	No
VDP LOGIC	"if-Then-Else" logic drives insertion of variable elements
WEB OUTPUT - PRINTING/URL/EMAIL	Yes- Cross media and Multi-touch Campaigns

SYSTEM REQUIREMENTS	FOR HTML E-MAIL OUTPUT
2 GHz Pentium processor, or faster	MS Internet Information Services v 5.0 or later
At least 512 MB RAM (1GB recommended)	MS SMTP Services
200 MB storage space	
True Color (24-bit) video card and display settings	

KODAK DARWIN VI AUTHORING TOOL

Unlike Pageflex, Kodak's Darwin VI Authoring Tool is compatible with both Windows and Macintosh platforms, as a plug-in to both QuarkXpress and Adobe InDesign. Though similar in many ways, Darwin also includes personalized URL tools built into the standard software.

Darwin offers many options to creating personalized documents, ranging from brochures and newsletters to more detailed options such as personalized information kits and datasheets. Darwin can accommodate large databases, allowing for personalization of color, text, graphics, and full background pages. There is also an option to include customized charts and/or barcodes within each piece.

The upgraded version of Darwin Pro can merge dynamic images with the text in an existing database. Personalized images can incorporate variable text data, making the resulting images extremely relevant and eye-catching.

Darwin has many other features that are similar to Pageflex. It also uses rule-based logic that allows for simple ways to integrate variable data into the document, as well as offering unlimited variability. Online demos are available for a customer to try out the product as a plug-in to an existing design software setup. Online "webinars" and training support are also offered.

APPENDIX

Table 5B.4 *Darwin VI authoring tool analysis*

ASSET FEATURES	Auto Text /Image fitting - Win version
COLOR MANAGEMENT	CREO Digital Color
COST OF PROGRAM	$2,995
DATABASE CLEANING	Yes - profile editing, upper/lower case changes, export database
DYNAMIC ELEMENTS	Text, Images, Color, Pages, Layouts
HARDWARE REQUIREMENTS -MAC	See below
HARDWARE REQUIREMENTS - WIN	See below
LANGUAGE SUPPORT	Supports non-Romance languages- Chinese, Japanese, Arabic, Hebrew etc.
MAIL FORMATS- DESIGN OPTIONS	Supports different page sizes etc, booklets
MERGE CAPABILITIES Y/N	Yes - profile editing, upper/lower case changes, export database
OUTPUT	CREO APR - reduces process time by allowing images to remain in central server
PLUG-IN INTERFACE	GUI
POSSIBLE COST ADDITIONS	VI Toolbox , Darwin Pro version
PREFLIGHT	Advanced Preflight - view specific record in real time
PROGRAM SUPPORTED - DESIGN SOFTWARE	InDesign CS2/CS3 and QuarkXPress
PROOFING CAPABILITY	Yes
RIP SUPPORT	Yes- RIPed once, CREO, Optimized PostScript, PPML, VDX, VIPP, PDF
SOFTWARE REQUIREMENTS	Adobe InDesign C2 or CS2 and QuarkXPress
SPECIAL FEATURES	Zipcode barcoding in Pro edition
VARIABILITY LIMIT	No
VDP LOGIC	Rule-based logic
WEB OUTPUT - PRINTING/URL/EMAIL	Yes

Database Technologies for Personalization

APPENDIX

MACINTOSH SYSTEM REQUIREMENTS	WINDOWS SYSTEM REQUIREMENTS
Power Macintosh G5 or INTEL-Based Mac	INTEL Premium processor
1G+ RAM	1 GB RAM
MAC OS X v10.4	500 MB storage space
500 MB storage space	

OBJECTIF LUNE'S PRINTSHOP MAIL

PrintShop Mail is available for both Windows and Macintosh platforms, and works as an enhancement to several design software applications. It is a stand-alone software program that can use design layouts created from various design software programs, including Adobe InDesign, Adobe Illustrator, Corel, Microsoft Word, and PageMaker. This allows a pre-created standard layout or mailing piece to be incorporated into the new software without having to start from scratch.

PrintShop Mail uses simple "drag and drop" technology to incorporate the variable data. Design options include image rotation, the ability to adjust for varying text and graphic elements, and the inclusion of bleeds and cropmarks. The software is compatible with most database formats and allows for instant previews of the dynamic content.

PrintShop Mail also offers a variety of upgrades and enhancements. For example, DirectSmile is a plug-in that allows for a limited amount of image personalization. But the graphics are very basic, and DirectSmile is limited compared to other programs.

Table 5B.5 *PrintShop Mail analysis*

ASSET FEATURES	Auto text/image fitting
COLOR MANAGEMENT	Version 6.1 offers enhanced color management and spot color
COST OF PROGRAM	$1,395
DATABASE CLEANING	Database Filter, compatible with almost any database format
DYNAMIC ELEMENTS	Text, Images, Color, Pages, Layouts
HARDWARE REQUIREMENTS -MAC	See Below
HARDWARE REQUIREMENTS - WIN	See Below

APPENDIX

LANGUAGE SUPPORT	Dutch, English, French, German, Italian, Japanese, Simplified/& Traditional Chinese, Spanish*
MAIL FORMATS- DESIGN OPTIONS	Direct Mail, brochures, catalog wrappers, personalized cards, web to print docs
MERGE CAPABILITIES Y/N	Yes
OUTPUT	Output to numerous printing formats to ensure printing at rated speed.
PLUG-IN INTERFACE	Direct Smile
POSSIBLE COST ADDITIONS	Direct Smile, PrintShop Web
PREFLIGHT	Pre-flight and real-time proofing of documents. Per record, per page.
PROGRAM SUPPORTED - DESIGN SOFTWARE	Print Shop Mail for Windows and Macintosh
PROOFING CAPABILITY	Softproofing for high or low resolution
RIP SUPPORT	Yes - Optimized PostScript, Fiery FreeForm, FreeForm2, Creo VPS, PPML, VIPP**
SOFTWARE REQUIREMENTS	Adobe Illustrator, InDesign, PageMaker, Corel, Microsoft Word
SPECIAL FEATURES	Barcoding, Built-in Acrobat Library
VARIABILITY LIMIT Y/N	No
VDP LOGIC	Drag and Drop, requires no programming
WEB OUTPUT - PRINTING/URL/EMAIL	Automatically generated web pages

Plus Turkish, Russian, & Portuguese for WIN
**Plus AHT (ColorFlare, One Rip) and PPML/VDX for WIN*

MACINTOSH SYSTEM REQUIREMENTS	WINDOWS SYSTEM REQUIREMENTS
Power PC Macintosh G3 or 350 Mhz or higher	Pentium II, 300 Mhz or higher
256 MB of RAM	128MB of RAM
50 MB of storage space	120 MB of storage space
	SVGA monitor resolution 800 X 600 high color (16-bit)
OS 10.4.9	Windows 2000, XP and 2003 Server

Database Technologies for Personalization

APPENDIX

XMPIE UDIRECT (CLASSIC, STUDIO, OR PREMIER)

uDirect by XMPie is a VDP plug-in that fits both Windows and Macintosh platforms, using Adobe InDesign as its design base and enabling a wide variety of VDP options. XMPie offers online demonstrations, tutorials and webinars for the product, which can be purchased, downloaded and installed through the XMPie online store.

uDirect Classic is the base VDP plug-in that works with Adobe InDesign to create personalized documents such as catalog wrappers, cards, books, brochures, and direct mail. uDirect Classic also allows for special text features such as text wrapping, adding shadows, and color scheme customization. Using the simple point-and-click technique with if/then logic, a static document can quickly become dynamic.

With uDirect Studio, the designer can add personalized images or graphics to an InDesign document by using links to Adobe Photoshop or Adobe Illustrator. With the uChart plug-in, uDirect Studio can create personalized, data driven charts. uDirect Premier has all the functionality of uDirect Classic and Studio, and can also use more complex logic to separate data from the design function and move it to a data administrator. XMPie also offers a plug-in called uImage that can add dynamic text and images as variable "smart objects" in the personalized document.

XMPie offers a tiered set of applications that can be used together, expandable to meet a company's growing needs. By upgrading to XMPie's PersonalEffect, a company can access all the functions from uDirect and also offer Web campaigns, response tracking, personalized Web sites and SMS messaging. Besides helping the user to create variable charts and images, XMPie add-ons also facilitate web-store functionality.

Table 5B.6 *XMPie uDirect Classic analysis*

ASSET FEATURES	Information not available
COLOR MANAGEMENT	Information not available
COST OF PROGRAM	$3,000
DATABASE CLEANING	No
DYNAMIC ELEMENTS	Text, images, color, pages, and layout
HARDWARE REQUIREMENTS -MAC	See below
HARDWARE REQUIREMENTS - WIN	See below
LANGUAGE SUPPORT	Information not available

APPENDIX

MAIL FORMATS-DESIGN OPTIONS	Information not available
MERGE CAPABILITIES Y/N	Yes
OUTPUT	PDF, PPML, PostScript, Kodak VDX, Creo VPS, Xerox VIPP
PLUG-IN INTERFACE	uImage
POSSIBLE COST ADDITIONS	uPlan, uCreate, uProduce, Chart, uDirect Studio, uDirect Premier, PersonalEffects
PREFLIGHT	with Adobe InDesign
PROGRAM SUPPORTED - DESIGN SOFTWARE	Adobe InDesign
PROOFING CAPABILITY	thru software program ie InDesign
RIP SUPPORT	with PersonalEffect and other upgrades
SOFTWARE REQUIREMENTS	Adobe InDesign, Photoshop, Illustrator, GoLive, Dreamweaver
SPECIAL FEATURES	Upgrades include uPlan, uCreate, uProduce, part of PersonalEffect
VARIABILITY LIMIT Y/N	No
VDP LOGIC	"if-Then-Else" logic drives insertion of variable elements
WEB OUTPUT - PRINTING/URL/EMAIL	Yes

WINDOWS SYSTEM REQUIREMENTS	MACINTOSH SYSTEM REQUIREMENTS	GENERAL REQUIREMENTS FOR BOTH
Intel Pentium 4 or better	PowerPC G4 800 MHz or better or Mac Intel	1 GB Ram (2GB recommended)
Windows XP with Service Pack 2	Operating System Mac OS X 10.4.8	80 GB storage (250 GB recommended)
		CD/DVD-ROM Drive
		10/100 Mbps Ethernet Adapter (recommended 1Gbps)
		Adobe InDesign CS2 or CS3

APPENDIX

NOTES

128. Raab, "The Market for Marketing Automation Systems."
129. Speier and Venkatesh, "The Hidden Minefields in the Adoption of Sales Force Automation Technologies."
130. Rigby, Reichheld and Schefter, "Avoid the Four Perils of CRM."
131. Jayachandran, Sharma, Kaufman, and Raman, "The Role of Relational Information Processes and Technology Use in Customer Relationship Management," p. 177.
132. Ibid.
133. Ibid.
134. Rigby and Ledingham, "CRM Done Right."
135. Ibid, p. 118.
136. CMO Council, "The Power of Personalization: The Impact and Influence of Individualized Content Delivery."
137. CMO Council, "The Power of Personalization: The Impact and Influence of Individualized Content Delivery."
138. Drozdenko and Drake, *Optimal Database Marketing*.
139. Ibid.

CHAPTER SIX

CORPORATE COMMUNICATIONS: IN-PLANT PRINT SHOPS AND TRANSPROMOTIONAL DOCUMENTS

WHILE THE EMPHASIS OF THIS BOOK is primarily on the marketing function of personalized communication, there is also a great deal of personalization being implemented through corporate communications to employees, stockholders, and influencers such as publishing media. Many large corporations and government entities maintain in-house printing services that print, distribute, and mail these documents. Corporations that provide financial and health care services also use in-plant services to communicate with customers. The first type of personalized corporate communication was transaction printing—the printing and mailing of statements, bills, and other documents to confirm or facilitate transactions between customers and the companies they do business with. Transaction documents were first printed digitally with early dot-matrix printing devices. While these mechanical printers printed variable-width typefaces, their low resolution limited their applications to office documents and business communications.[140]

This chapter examines the corporate in-plant printing function and the new opportunities it offers for growth by combining transactional documents with promotional messages.

IN-PLANT PRINTERS

A 2007 InfoTrends study estimated that 70% of companies with more than 1,000 employees have in-plant print shops.[141] Moreover, over 60% of those surveyed expected that the volume of printed material produced by their in-plants would increase in the near future.[142] The largest in-plant print shops are located in government facilities, universities, insurance and financial firms, retailers, and health care companies.[143] The U.S. Government Printing Office is the nation's largest in-plant, serving as publisher for the federal government. Its mandate is to distribute information products and services for all three branches of the federal government, and in this capacity it produces the Congressional Record and all other legislative information supporting the U.S. Senate and House of Representatives.

In-plants in corporate settings such as utility companies, insurance and investment firms, and telecommunications firms print the transaction records and bills that are the heart of their customer interaction. In addition to these transaction documents, in-plants also print marketing brochures, newsletters, signs, manuals, reports, and catalogs. In-plants also provide a range of document services, including mailing, copier management (for printers distributed across multiple departments), and scanning for archival purposes.

In-plant printers can become the internal resource for a company's personalized direct mail, aimed at acquiring or keeping customers, by applying principles from the first five chapters of this book. The primary challenge for the in-plant to become the "vendor of choice" for a company's marketing communications tactics is to provide service that exceeds that of the commercial printers, mailers, or service bureaus that marketing decision makers currently use. This internal marketing challenge is similar to that of the commercial printer: to get a seat at the table with the marketing managers as they plan their communications strategies. The case study presented below describes how the RIT in-plant printing services center evolved to become a marketing partner to the university's admissions process. To transition from being "just a printer" to becoming a major division's marketing partner, this in-plant followed five discrete steps:

- Get to know the marketing decision makers of the firm and understand their customer engagement strategy.
- Start small with one program and measure the results.
- Get it right.
- Build additional programs with that internal customer.
- Sell to other internal customers.

EVOLUTION OF AN IN-PLANT: A CASE STUDY OF PRINTING SERVICES AT RIT FROM 2005 TO 2008

In 2005, the Printing Industry Center at RIT conducted three case studies to describe current in-plant printing challenges and opportunities.[144] We asked where printing was done within each enterprise and where the organization sourced its other printing work. The cases included one university (RIT), one large retailer, and a food packager. This current study returns to RIT to determine how things have changed in the intervening three years.

CORPORATE PRINTING AT RIT IN 2005

In the spring of 2005, the Rochester Institute of Technology, an upstate New York private comprehensive university, had approximately 15,000 students, 700 faculty and 2000 staff members. Printing capabilities were spread throughout the organization: on the desktops of faculty and staff members, via networked printers in teaching labs, and through a centralized in-plant print shop that had been in existence since 1972.

RIT's printing needs can be classified into two major categories:
- print that serves primarily an internal audience, including both administrative and educational community members, and
- print that reaches an external audience for the purpose of marketing and recruiting.

The 2005 study did not cover the printing activities of RIT's School of Print Media to deliver its degree programs, the RIT Printing Applications Laboratory that conducts research and testing on a propriety basis for external clients (such as paper and equipment manufacturers), or the RIT Cary Graphic Arts Press.

PRINTING FOR RIT'S INTERNAL AUDIENCES

Print to support the internal needs of the organization was (and is still) managed by three types of administrative units on campus.

- The Office of Finance and Administration manages the in-plant print shop, copy center and mailing facility. This office also manages procurement, overseeing the purchase of high-end networked printers throughout the university.
- The staff of the Chief Information Officer manages both RIT's Data Center and a few of the institute's computing labs used for teaching.

CASE STUDY

- College and administrative departments purchase desktop print devices for faculty and staff, networked printers for the student labs they manage, and large photocopiers to do departmental copying.

RIT Office of Finance and Administration

We interviewed Jim Fisher, at that time the assistant vice president of Finance and Administration, who manages both in-plant print shops on campus and the procurement process for the entire university. He first explained his role as procurement manager. Equipment purchases over $1,500 must involve RIT's Central Purchasing office. According to Fisher, there were 166 networked printers over this price range on campus in 2005. The manufacturers of these printers were limited by RIT's preferred vendor program to include Lanier (n=11), Konica (n=55), Canon (n=43), and Xerox (n=57) printers. RIT acquires approximately 20-25 networked printers per year throughout the campus. Most are replacement copiers, but an additional 1-3 printers per year are purchased. Whether and how involved the procurement office gets with these purchases is usually determined by each department, as long as it chooses a preferred vendor or vendors. Some departments work with Central Purchasing from the start, while others want help to make the final decision. Still others will select the make and model that meets their needs and engage Central Purchasing only in the last stage of placing the order. Out of these three scenarios, the first is the most widely used.

Does RIT have a document management strategy? For the most part—no. However, in 2004, Xerox became the preferred vendor for printing and document management needs.

RIT's In-Plant Printing

Jim Fisher also supervises the RIT in-plant printing facility, which consists of two major service centers. The older one is the HUB: Central Print & Postal Services (commonly known as the Central HUB), located on the outskirts of the campus. This facility houses a print shop, copy center, and mailing facility. In 2005 the print shop had three printing presses: an AB Dick 9850 (purchased in 1988), a Heidelberg single-color Quickmaster (purchased in 1998), and the Heidelberg Speedmaster 74 (purchased in 2002). The Central HUB copy center housed a Xerox Docutech and an iGen3, the latter installed in spring 2005. The Central HUB facility serves the printing needs of the RIT administrative units, including the offices of alumni relations, admissions, and government affairs, and the president's office. Its print shop staff had six full-time employees in 2005.

CASE STUDY

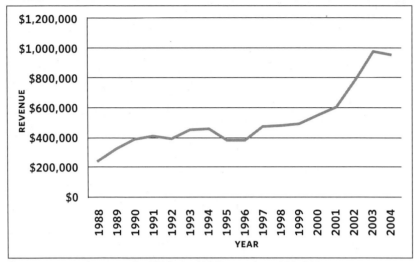

Figure 6.1 *Annual revenues for RIT Central HUB print shop, 1988–2004*

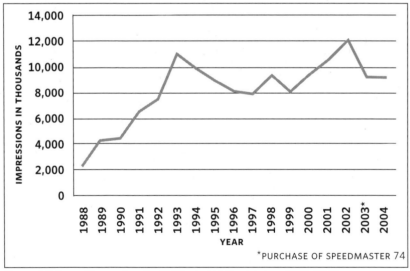

Figure 6.2 *Annual impressions for RIT Central HUB print shop, 1988–2004*

Print volumes for a recent 17-year period for just the print shop portion of the Central HUB facility are presented in Figures 6.1 and 6.2. Print revenues (internal charge-backs) increased steadily over this period, with a sharp rise in 2003, the year the Heidelberg Speedmaster was in full operation, which allowed the center to complete four-color print jobs in-house. Because the Speedmaster could handle a wider paper, the number of impressions de-

CASE STUDY

clined (due to the ability to print two-up and four-up on a page). However, all of the Central HUB's presses remain active, including the AB Dick press which reportedly is running every day.

The Central HUB copy center's new iGen3 replaced one of the two high-volume monochrome copiers that had been there since 2000. The volume for the Docutech has declined, due to the spread of desktop and networked printers throughout the campus. RIT contracted with Xerox for an operator for the iGen3 for three years.

The Central HUB print shop also serves paying customers in the community. As of the 2005 study, this revenue accounted for about 5% of the shop's annual total. Outside clients are small businesses and not-for-profit firms in the area.

The second print facility on campus, Crossroads HUB, is a "Kinko-like" retail storefront located in the Crossroads student center. This facility is managed separately, but also reports to the Director of Printing and Mailing Services (under Fisher). It was created in 2000 and houses a Xerox 6060, a Neuvera 120, an HP 60-inch, wide-format printer (the busiest printing device in the shop), and a number of PC and Mac workstations for file preparation. A high volume of jobs is submitted by students. The key challenge is to make money on short-run jobs, the majority of which need a high degree of file correction and/or preparation. Much staff time is devoted to working with naïve users who do not submit the right file formats to ensure top-quality output.

Print revenues for 2003–2004 for both the Central HUB and Crossroads HUB centers are summarized in Figure 6.3. Internal charge-backs (services charged to RIT departments) decreased during that period, but the revenue from cash transactions at the Crossroads copy center increased.

Data Center

The RIT Data Center is managed by Debra Fitts, who reports directly to the director of Technical Support Services, an administrative unit that is the responsibility of RIT's CIO. The RIT Data Center prints for the admissions, housing, registrar, bursar and payroll departments on campus. All of the print jobs are generated from internal RIT student and administrative databases. The storage and maintenance of the databases is managed by an outsourced service vendor. As of 2005, there were five staff members in the Data Center, including two press operators, one for each shift. This number reflected a staff reduction, due to the outsourcing of programming and database administration to vendors. Retired workers were not being replaced because of productivity improvements through automation. The Data Center's biggest

CASE STUDY

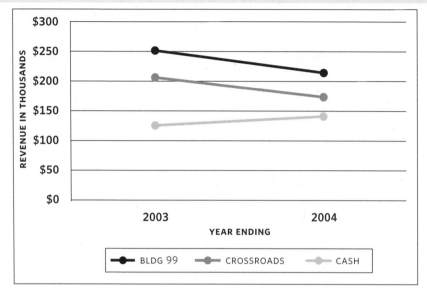

Figure 6.3 *Revenues from RIT copy centers (charge-backs for both Central and Crossroads HUBs, and cash transactions from Crossroads HUB)*

need was for scheduling software to keep track of cascading jobs.

The amount of Data Center printing declined substantially in the decade previous to 2005. At one time, the Data Center printed about 500,000 pages per month, but by then was producing only about 200,000 pages per month. There was an ongoing debate about whether the Data Center "should be in the printing business." For example, Fitts learned that RIT was one of the few educational data centers that were still printing internally. The volume of printed pages coming out of the Data Center had declined primarily because of the installation of an Oracle database system that manages HR, payroll, and the internal accounting systems that distribute data files electronically throughout the campus to key budget managers. The monthly operations budget statements for each department are now printed on the desktop of the accounting manager in each college and department. Before the Oracle system was in place, reports were printed centrally and then mailed to each department manager. For the HR function, payroll deposit statements are now provided only in electronic form; if an employee wants a paper copy, he or she can print it on the desktop.

As of 2005, the following documents were printed by the Data Center for internal RIT departments. With very few exceptions, all were self-mailers. The folding was also provided by the Data Center.

Corporate Communications

CASE STUDY

- **Admissions:** reports, letters to applicants
- **Financial aid:** award letters (annually), amended award letters (quarterly), promissory notes (on pre-printed shells)
- **Housing :** confirmation letters
- **Registrar:** "21 day" reports (quarterly), class lists (quarterly for over 400 classes offered), grade reports, schedules of classes
- **Bursar:** quarterly bills. (In 2005, RIT began to offer paper billing only by request. All others received e-billing.)
- **Payroll and accounting:** student checks (on a desktop Lexmark printer), accounts payable, and W2 forms (which were slated to migrate to the desktop)

Departmental Printing to Support the Core Mission of RIT

RIT's academic departments are organized into eight colleges. The E. Philip Saunders College of Business is used here as an example to assess the amount and nature of the printing in the college units. Printing within the colleges at RIT can be placed into three categories:

- printing in support of faculty teaching and research,
- printing in the computer labs (by students), and
- printing for the administrative unit.

The College of Business also contracts with external print services providers to produce its glossy sales brochure, called a "Viewbook" (further discussed in the next section).

It is the responsibility of Dave Ballard, systems administrator, to purchase and maintain the computing and printing equipment and manage the print capability of the College of Business. As of 2005, there were 58 printing devices in the college, for a faculty and staff of approximately 60 people. Nearly every faculty and staff member had an office desktop printer of some vintage in his or her office. To support the college's administration, there were ten networked printing devices (two of them color devices) to serve seven administrative offices. In addition, two black-and-white copiers directly supported the copy needs of faculty members for their classes (a Xerox Workcenter Pro 90 and a Xerox 460DC). Print volumes on these copiers averaged approximately 100,000 pages per month during the school year. Typical documents copied on these copiers were exams, syllabi, course handouts, and presentation materials.

There were two networked printers in the College of Business student labs, both Dell 5300 Ns, purchased in 2004. The four-year service and warranty that was included in the price was a key purchase criterion. Print jobs are submitted by students who are completing assignments for their classes, and there is also an unmeasured amount of printing for personal reasons. Ballard estimated in 2005 that 250,000 pages per year were printed in the labs, and this figure had been increasing annually. No one knows exactly how much is printed because the systems administrator tracks printing volumes primarily by the amount of paper purchased. According to Ballard, the biggest challenge was managing print volumes in these teaching laboratories. There were print policy guidelines for the students, upheld by lab assistants who reviewed and deleted inappropriate jobs in the print queue (e.g., no PowerPoint slides were allowed, no pdf files, and no documents over a certain megabyte limit). But there was no limit to the amount of sheets students could print within any one quarter. The lab managers reviewed Pharos software to monitor and control printing costs. This new software would allow for rules to govern the number of sheets printed, and to charge student debit accounts if a certain limit was exceeded. The software was not purchased due to its high cost.

PRINTING FOR EXTERNAL AUDIENCES—MARKETING AND RECRUITING

We also interviewed Dan Shelley, the director of undergraduate admissions at RIT since 1990. It is Shelley's job to deliver the entering class of undergraduate students each year. Details regarding the precise printed publications he uses were obtained in a separate interview with Bob French, at that time the assistant vice president of enrollment and career services. French manages the University Publications department, responsible for creating and printing RIT's recruitment publications.

The undergraduate recruitment effort consists of an integrated print and web campaign in addition to a number of on-campus, face-to-face meetings with prospective students and their parents. In 1990, only print and phone were used for marketing communications activities. By 2005, the media mix included print, phone, Internet, and email activities. Print budgets remained steady as students and parents continue to demand tangible promotional materials, but a decline in the cost of print allowed RIT to print a larger quantity and wider variety of materials. In 1990, the primary printed materials were Viewbooks for each of the eight colleges (glossy, 32-page publications), a course bulletin (400 pages printed on uncoated paper), and personalized letters. RIT outsourced the printing of the Viewbooks and bulletins to external printing vendors. Additional printed items were application forms (monochrome) and maps (color). Individualized letters to prospective

CASE STUDY

students were and still are printed internally. The amount of internal printing (personalized letters, application forms, and related responses) increased, driven by the number of inquiries. In 2005, the University Publications department's printing devices included an HP 5 SI and an HP 4350 TN. The Admissions Office also had an HP 4500 color laser printer for campus maps, etc., and an HP 4050 for other printed materials, such as packets for visitors at open houses.

But already by 2005, printed publications for admissions purposes had been dwarfed by Internet and email recruitment. Electronic marketing efforts were added without hiring more staff members. In general, software efficiency was a key driver in allowing staffing levels to remain the same. While Web maintenance for the undergraduate admissions site was once outsourced to another division on campus, in 2005 the Admissions Office had a team of four individuals to do it internally. To augment phone interactions, the Admissions Office also had full-time staff members devoted to Web chat inquiries.

Electronic document management was in place in 1999 with the use of a DocImage system for transfer student evaluation, where the transcripts of applicants need to be shared among a number of different academic units on campus to award transfer credit. Most paper documents coming into undergraduate admissions, financial aid and some other offices were scanned, stored, and destroyed.

Printed recruitment materials were typically outsourced to commercial printers, with a limited volume going to the RIT in-plant print shop. Two local Rochester printers, Tucker Printers and Mercury Print Productions, Inc., provided these materials, although a competitive bidding process engaged a number of other printing firms. For the approximately 25,000 copies of the annual course bulletin, a web press vendor in a nearby state had received the print contract for a number of years. In 2005, the University Publications department had a half-time print buyer to help in the purchasing process. However, it was predicted that a one-vendor model might be implemented because of the extra cost charges that appear with overruns and changes in copy, items that are normally associated with serving the customer in print buying.

UNIVERSITY NEWS

The other RIT department with heavy publishing needs is University News. This division is headed by Bob Finnerty, and is responsible for RIT's two publications for external constituents: a tabloid-style newsletter titled *News and Events*, and the *University Magazine*.

The 2005 circulation of *News & Events*, printed by an outside vendor, was about 6,000. (Print providers were changed often based on price. In 2005, the printer for this job was Rochester's Microera Printers, Inc.) *News & Events* is distributed to all faculty and staff members through the internal RIT mail system, to students and visitors through campus information boxes, and by mail to retirees, friends of RIT, politicians, and the media. The circulation grew in the early 2000s by approximately 500 simply because of RIT's growth. The budget also increased slightly because of the circulation increase, the rising cost of newsprint, and an expanded use of color. Since the fall of 2004, *News & Events* has been printed in color on all pages and in every issue. In addition to print, *News and Events* has a strong Web presence at www.rit.edu/newsevents. In 2005 there were approximately 35,000 page views per month on this site alone, and in a typical month, about 500 documents were downloaded from the site.

The University Magazine was created in 1999. As of 2005, faculty and staff, parents of current students, and friends of RIT received it three times per year, twice with 48 pages and once with 32 text pages and a 70-page donor report. The magazine was printed by web offset at the Lane Press in Burlington, Vermont, which was also responsible for its mailing. Circulation grew from 92,000 to 115,000 in its first six years, and RIT budgeted about $300,000 per year for the magazine. While RIT was able to realize a cost savings in postage, a dilemma was brewing: As the alumni base keeps growing, how will RIT maintain the growth in circulation for the same budget? The magazine also has a Web version at www.rit.edu/magazine. As of 2005, it got less Web traffic than *News and Events*—about 6,000 page views per month. Additional content is added on the magazine's Web site, such as audio/slide show stories.

2005 STATUS OF RIT IN-PLANT PRINTING

The amount of internal printing at RIT increased between 1990 and 2005, as evidenced by an increase in printing at both HUB locations, increases in the number of networked printers purchased, and the increase in the quantity and variety of printed publications for admissions. The only RIT areas that reported decreases in printing were the Data Center (printed forms) and the Central HUB (also printed forms). In the Data Center, electronic distribution of content had eliminated the need for printing and distributing statements, bills, letters, and financial information for faculty, staff and students. As for the Central HUB, the increase in the number and capabilities of networked copiers and printers proliferating throughout the campus and the electronic distribution of classroom materials may explain its decrease in printing.

For RIT's external audience, however, electronic distribution of recruiting materials has not entirely replaced print. Print budgets remained relatively constant, even though electronic means of marketing were being used to enhance contact with external audiences and prospective students. As of 2005, the Admissions Office still used external print providers to print large-run, perfect-bound materials, and continued to print its own personalized letters to students. It also used the HUB for some marketing materials. The president's office still used the HUB to print programs for the multitude of honors and awards ceremonies each spring, including a graduation booklet that lists the names of all graduating students.

While in-sourcing of print is still high at RIT, the locus of that printing for internal audiences changed from the centralized copy machine (in the department or one of the in-plant copy shops) to the desktop. For example, in the student computer labs, electronic access and distribution of course materials (e.g., Internet access to library databases) increased the amount of material printed by networked printers in the labs. In addition, a Web-based course management information system deployed throughout the campus enabled faculty members to post materials that were previously printed and passed out to students in class. In many cases, faculty members distribute course content via the course Web site, and students have the option of printing it out.

Although we did not have an exact measure of how much material was electronically distributed as of 2005, the Central HUB reported that the number of requests for "course packs" had declined. (Course packs are bound volumes of reading materials gathered by faculty members from a variety of sources, with copyrights obtained, copied by HUB and available for purchase in the RIT bookstore. The reduction of these "do-it-yourself" course packs was facilitated by major textbook publishers who have created elaborate custom Web sites to provide supplemental readings and other activities for instructors who adopt their textbooks.)

THE RIT IN-PLANT IN 2008

One major change that impacted RIT's in-plant printing was the hiring of a new manager, John Meyer, in early 2006. Meyer came to RIT from a local Rochester print services provider where he had served as the director of operations for five years. After settling into the job, he identified two strategic initiatives to improve services. The first was an electronic job submission system, that would allow RIT departments to access printing services in a more timely and convenient fashion. The other idea was to add variable data printing (VDP) services to expand the volume of print for the iGen3. Since

CASE STUDY

both initiatives required capital investment, Meyer chose to start with only one of them, VDP. In the summer of 2007 he purchased XMPie software and trained his staff and RIT University Publications designers to use it. By fall they were ready to offer VDP capabilities to all RIT departments for their communications needs.

Meyer's first customer was the Admissions Office. A variable data postcard campaign for the office had been initiated during the previous year (see Figure 6.4). The idea for the campaign was developed by two students in the School of Print Media in 2005, Adam Peck and Nick Bradish, as a service project for *PUB*, the school's print media club. The students had presented their idea to Dan Shelley, who immediately saw its promise as a creative way to get RIT noticed in the clutter of mail that fills any college applicant's mailbox.

The students had developed a simple software algorithm to calculate the mileage from the home address of an RIT applicant to the RIT campus, using a typical online mapping program such as Mapquest. This mileage data was appended to files containing applicants' mailing addresses, intended programs of study, and names of the high schools they were currently attending. The data was integrated with the design, and PDF files of the postcards were then delivered to the Central HUB for printing and mailing.

Figure 6.4 *Variable data postcard developed by RIT students*

Because of the success of this campaign, the Admissions Office wanted to work with the Central HUB again to develop a new VDP application to target

Corporate Communications

CASE STUDY

prospective students earlier in the admissions process. The Admissions Office had been approached by an external digital printing service that offered a complete turn-key program to create a partially-completed application to RIT with known information about a prospect obtained through a list provider. However, the office decided to give Central HUB an opportunity to match this external offer, using its recently-acquired design capabilities and XMPie expertise. The job was named the "preferred application project," and Admissions gave the HUB five weeks to design, print, and mail over 30,000 of these applications to a pool of highly-qualified students across the U.S. Figure 6.5 presents an example of the primary piece in the mailing.

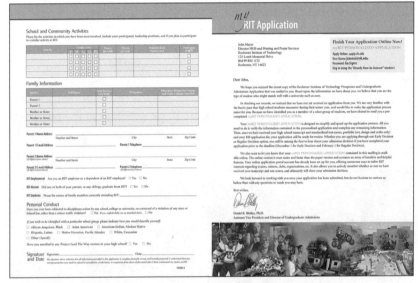

Figure 6.5 *An example of the "Preferred Application" variable data mailer*

In order to meet its deadline, the Central HUB staff worked with designers in RIT's University Publications office and database experts at RIT's Information Technology Services office. The design challenges included those typically found in primarily text VDP applications, such as allowing enough space for the very long surnames of a few applicants. Though the piece was printed in color on the iGen3, there were no variable images on it. In addition to the application form, the Central HUB also printed the additional items for the mailing—envelopes, a list of programs at RIT, transfer course information, and a page for the applicants' written personal statements.

Two weeks after the package was mailed, the Admissions Office saw a 35% increase in applications compared to the same time period during the previous year. Bravo! Because of projects such as this one, the volume of printing on the iGen3 has nearly doubled in the time that Meyer has served as director of HUB. He is using these successes to make the case for VDP at other RIT administrative offices. Meyer has approached both the Alumni Relations and Development offices with ideas for mailings.

Although the Central HUB has seen the volume on its iGen3 increase, it has also witnessed a decrease in the demand for previously printed communications. For example, the Admissions Office now uses email more frequently than printed mail to reach some prospects. While the number of its printed mailings has gone down, the ones that have been retained are more valuable, with demonstrably high impacts, as noted above. The HUB now mails a five-color offset-printed self-mailer that has been outsourced to an outside print services provider.

Other changes are occurring for RIT's Data Center. By working in closer cooperation with the HUB, the Data Center has dropped the lease on its monochrome copier and is now using printing services at the Crossroads HUB for selected documents. Currently, the Data Center's monthly page count averages 120,000 pages of black-and-white output. In addition, the Data Center is continuing its migration to electronic distribution of important internal documents, such as quarterly grade reports for students.

The remainder of this chapter explores two other growth opportunities for corporate in-plant printers: document management (including content management and archiving) and transpromotional printing.

DOCUMENT MANAGEMENT

Document management is the "generation, archival, retrieval, dissemination, and termination of documents as they initially exist."[145] *Content management* decouples the information within a document from its original form and, through the use of metadata and tags, allows this information to be stored, retrieved and published (sometimes automatically) in another form. For example, a product description that originally appeared in a printed brochure as sales collateral could be tagged in XML, stored in a digital asset management system, and retrieved to be published on the corporate Web site's e-commerce pages.

Content and document management functions have become increasingly important to companies as they attempt to manage the accuracy and consistency of information published across a wide-range of media platforms. This multi-channel communications challenge was documented in a 2006 study by InfoTrends. The study identified the nature, purpose and amount of multichannel communications within firms, and the challenges they face in managing the workflow.[146] In the study, over 300 IT and business professionals who were knowledgeable about their firms' multichannel communications or content management functions were asked to estimate the percentage of content that is delivered across many media channels. The study found that over 50% of all content for product support, marketing, sales, and employee communication required multichannel distribution. The most popular media channels were email (58.4%), print (51.4%), internally-facing Web sites (48.1%), and externally-facing Web sites (40.4%). When asked whether their firms had a content management strategy, 36% of respondents replied yes, for the entire enterprise, and 33% replied that a division or department of their companies had one. The remaining 31% said no or were not sure. However, half of those with a positive response were only in the early stages of planning and developing the strategy.

What role does the in-plant print shop play? In a 2007 study of in-plant printers published by EDSF (the Electronic Document Systems Foundation), Ken Macro and his students at the California Polytechnic State University (Cal Poly) surveyed 80 in-plant printers (primarily in educational institutions) on their knowledge and current participation in their firms' content and electronic document management efforts. Macro found that nearly half

(45%) of the in-plants had a document or content management initiative in the works. The top electronic services these in-plants offered were: print-on-demand (91%), digital archiving (51%), and electronic file transfer protocol (34%). Fewer offered database management (15%), content management (9%) and user access to warehoused files (9%).[147]

However, only 27% indicated that an in-plant printing operation should be the initiator of a document management or content management system for its organizations. The majority believed that such a plan should be initiated by the company's IT department.

The limited role that in-plant printers have in document management is not surprising, according to the InfoTrends study. Macro and his team interviewed primarily IT professionals and noted that IT divisions are the primary developers of firms' content management systems. Macro wrote, "In-plants must develop a cohesive and collaborative bond with IT services, data centers, and records management personnel in order to politically align the in-plant with the core mission of the parent organization..."[148] He goes on to recommend that the in-plant print facility of the future should place more emphasis on file storage and content management. While an in-plant shop may not lead the firm's document management strategy, it can be a key player in its implementation.

TRANSPROMOTIONAL PRINTING

According to the U.S. Postal System's 2007 *Household Diary Study*, U.S. households received 100 billion pieces of mail in 2007.[149] This figure includes over 18 billion bills, over 7 billion statements, and over 733 million requests for donations from charitable organizations. Banks, credit card companies, brokerage firms, retail companies (with their own credit cards), and insurance, utility, cable and phone companies were the primary sources of this mail. As of 2006, there were over 114 million U.S. households, an increase of approximately one million from 2005. While the number of households increased, the amount of transaction mail remained roughly the same from 2005 (40,990 million pieces) to 2007 (41.512 million)—only a 1.3% increase. The *Household Diary Study* attributes this trend to the growth of electronic billing and payment.

Electronic diversion continues to erode the volume of mail payments in favor of online payments, automatic deductions from bank accounts and other electronic methods of bill payment. As a result, the share of bills paid by mail dropped from 25.3% of total mail transactions in 2005 to 23.6% in 2007.[150]

One growing area in the printing of statements and bills is to add marketing messages into these transactional documents. The term "transpromo" (short for transpromotional) has been coined to describe this new way to reach a firm's customers—placing promotional messages on the statements in mailed documents that customers almost always open and read.

The inclusion of promotional messages in billing and statement mailings is not new. For years, utilities, telephone companies, and retailers have used inserts to cross-sell and upsell their own or other firm's products and services. But lately, the interest in transpromotional documents has been growing for three reasons:

- customers' positive attitudes towards mail as their preferred medium for promotional messages,
- advances in computer, database and printing technologies, and
- increasing postal rates.[151]

Let's examine each one of these, starting with consumer preferences for advertising media. With the proliferation of new media, consumers have been enjoying the benefit of boundless two-way communication opportunities. Advertisers are eager to take advantage of these media as new ways to reach consumers, especially young, mobile consumers. However, consumers are not particularly enthused about these primarily personal communications media using commercial messages. As noted in the first chapter, Facebook encountered a virulent consumer backlash when it attempted to include ad messages for products mentioned by members within their pages.[152] Heavy email users are also intolerant of the proliferation of SPAM overwhelming their accounts. The InfoTrends 2006 *Future of Mail* study documented people's preferences for how commercial messages reached them from the companies with which they do business. Direct mail was preferred by 61% of the respondents, email by 21%, telephone calls by 6%, and cell phone communications by 3%.[153] A Printing Industry Center research study had come to similar conclusions in 2002, when over 82% of respondents reported that they liked to receive catalogs from stores they patronize. Slightly less than half (49%) of respondents preferred email messages versus postal mail from companies they do business with.[154] Upon further analysis, we found age differences in the preferred media in the expected direction—younger people reported a higher preference for email than for postal mail. In the InfoTrends study, 58% of respondents said that companies use email to contact them regarding important information about their accounts. Those with Internet access also understood the advantages of online bill presentment services.[155] In general, consumer preferences vary by the nature of the information (e.g., email on fraudulent account activity) and by the consumer's comfort with electronic communication technologies.

Notwithstanding generational differences, mail from the U.S. Postal Service is accepted by the majority of adults as a preferred advertising medium. The inclusion of relevant marketing messages on important printed documents is an increasingly viable way to make current customers aware of new products or services from the businesses they work with. Consumer acceptance, though, depends on whether the messages are considered relevant and address past or future needs. The same InfoTrends study found that over half of the respondents (56%) indicated a preference for personalized offers that are designed around their needs and interests.[156] This also means that just under half of the respondents do NOT prefer such personalized offers, a fact that should warn firms that have a great deal of personal information about their customers—such as banks, investments firms, or health care companies—to carefully assess the preferences of their customers towards personalized communications.

Our ability to personalize messages has to be tempered by the desires of our customers. For example, in a recent study commissioned by the Print Industries Market Information and Research Organization (PRIMIR), over 500 U.S. adults were asked what they thought about promotional printing appearing directly on their bills and statements. A large majority (72%) said that this was not ideal.[157] One reason given for not welcoming promotions printed directly on transactional documents was that offers printed "directly on a statement" made that statement more difficult to read. Just because we can personalize messages does not mean we should do so in all cases. We should give customers the power to opt-out of these communications if they so choose.

The second reason why transpromotional documents are increasing in popularity is because of advances in technology, particularly in the speed and quality of digital color production printing devices. New printing presses by Kodak (Versamark), Ricoh/IBM (InfoPrint), and Océ (Jetstream) can print up to 1,000 color pages per minute. While most transactional documents are still printed monochromatically, InfoTrends found that 23% of the total transaction document volume was printed in full digital color in 2005. This was more than double the 2002 volume, when an estimated 11% of transaction documents were printed in color using digital technology.[158] This is an amazing increase given that it is much more expensive to print digitally in color. InfoTrends estimated that it costs 6.5 cents more to print a four-page document with two pages in full color than to use pre-printed color forms.[159] Even though the cost to print in color is declining and color quality is improving, it is still more expensive to use color in transactional documents.

A key technology facilitator beyond the printing press is *electronic verification technology*. These systems use in-line cameras to scan and monitor that documents are assembled correctly—in other words, that the information for a certain customer is inserted into the appropriate fulfillment package. Another new technology integrates the manufacture of document and envelope in one step, eliminating the need to check whether documents are assembled correctly.[160]

The third reason why transpromo documents are becoming ubiquitous is the problem of ever-increasing postal rates. Transpromotional capabilities provide motivation for banks, health care companies, investment firms, and utility companies to redesign their statements based on the information needs of their customers.

Where to Start

The impetus to consider using transpromo tactics often begins with the need to redesign a statement. A typical example of this is when businesses take on a new corporate identity. This was the case with Ameriprise Financial, as shown in the following case study, published by PODi.

CASE STUDY: AMERIPRISE FINANCIAL STATEMENT REDESIGN GENERATES ACCOLADES

by PODi (Print on Demand Initiative)
Excerpted from *Best Practices in Digital Print 2007*
Reprinted by permission of PODi (www.podi.org)

Ameriprise Financial's statement redesign enabled the company to reduce the amount of paper used and made it easier for clients to understand their financial position. Variable data text blocks also allowed for cross-sell and up-sell opportunities.

PROGRAM OBJECTIVES

- Redesign statement to make it easier to understand
- Reduce the impact the redesign has on production workflow

SIGNIFICANT RESULTS REPORTED BY USER

- Number of pages needed for some statements reduced
- Customers and advisers like the new design
- Variable data text messages can be incorporated into statements
- Information not applicable to a specific recipient can be omitted
- New workflow did not require substantial changes on data processing back-end

DESCRIPTION

When American Financial spun off its American Financial Adviser division, it got a new name—Ameriprise Financial. Along with the name change came new logos and opened the door for a redesign of the financial statements. Working with Art Plus Technology, the company changed the format of the statements.

The previous design was arranged by category or type of account, which made it difficult for clients to scan the statement and get a quick snapshot view of the overall holdings. In addition, each fund or product in a statement had its own page and there was a lot of empty space on the pages.

James Fenter, Technology Project Manager at Art Plus Technology, recalls that the mail pieces were heavy and contained many disclosure statements, which were not applicable to every client. Making sure that a particular client would only get the disclosure information that applied to that person was one of the first and biggest changes they made.

After doing some research to determine what the clients and finan-

CASE STUDY

cial advisers wanted to see in the statement, the redesign process advanced quickly. The new statement presents a quick overview of the customer's portfolio and what has happened since the last statement. A growth chart appears on the first page and asset allocations are provided for the current period and last period, and a 12-month recap is outlined and displayed graphically in the form of a pie chart. Another pie chart summarizes the Brokerage Portfolio Allocation section and insurance and Annuity products are summarized as well.

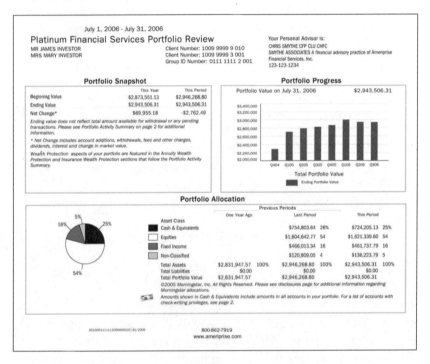

Variable text message blocks are sprinkled throughout the statement, allowing Ameriprise to mention specific products with the regulatory messages relevant to that product. The pension statements production process has been streamlined to allow some statements to be printed in duplex mode. Formerly pension statements had to be separated from the rest of the run and printed in simplex mode. The improvement reduced the number of sheets printed for the pension statements.

Adding a barcode to the address section has helped Ameriprise track returns for address corrections without needing to open the envelopes. In addition to printed output, statements are available online using XTND and XNET

CASE STUDY

to create PDFs that can be viewed by clients, advisers, and Ameriprise staff.

Fenter says reaction from clients and advisers has been "overwhelmingly positive." They find it easier to understand the information presented, plus it allows Ameriprise staff and advisers to take advantage of up-sell and cross-sell opportunities. Plus, Art Plus Technology was able to create a custom software solution that connected the Ameriprise Financial mainframe to the statement production solution. Thus, the company was able to keep its existing printing equipment and did not need to make major changes to its data structures.

CLIENT	Ameriprise Financial, Inc. http://www.ameriprise.com Ameriprise Financial, Inc. is a leading financial planning and services company with more than 12,000 financial advisors and registered representatives.
PRINTER	In-plant
AGENCY	Art Plus Technology http://www.artplustechnology.com Art Plus Technology is a Boston-based firm that specializes in designing and implementing complex health care and financial transactional communications for its clients.
HARDWARE	4 of the Océ Vario-Stream 7650 Finishing: Pitney Bowes - 2 - APS, 4 Series 8
SOFTWARE	Exstream Dialogue, custom software developed by agency
TARGET AUDIENCE	Ameriprise clients and financial advisers
DISTRIBUTION	900,000 monthly
DATE	August 2006, ongoing

For Ameriprise, all marketing materials, signage, stationery, business cards, and statements had to be redesigned. The company brought in the consulting firm Art Plus Technology to create the new look. In an interview published by InfoTrends, the founder of Art Plus Technology, Elizabeth Gooding, stated:

> If you consider the statement to be the underlying vehicle for a new type of marketing platform, you need a good design and structure to support messaging. Companies need practical advice on how to determine what to promote to each type of customer and the best way to implement these promotions on statements. More importantly, they need to take a critical look at their entire statement process.[161]

Gooding pointed out that even a fabulous statement design can be a waste of money if the right process is not used to push it through compliance and follow up on promotions.[162] She explained that the challenge companies face in implementing transpromo is determining how much personalization they want to use, which groups to focus on (the classic problems of segmentation), and where you put your money. Oftentimes you can save money and even generate incremental revenues by redesigning the statement, but someone has to develop the content and manage the process, and that is where these projects sometimes fail.[163]

Gooding added that service providers often paint a very optimistic picture, and customers are not always informed about the true level of effort that is required:

> We do customers and the industry a disservice if we don't paint an accurate picture... If [this strategy] was a walk in the park, everyone would have been doing it years ago. It takes work, collaboration, discipline, and a cross-functional team.[164]

Another impetus for moving into transpromotional printing is the upgrading of printing presses in a data center. The technology is now in place to support new ways to communicate with customers through the printed statement. Firms that are automatically poised to make the initial move into transpromotional printing are financial printing service bureaus. The following case study (excerpted from the InfoTrends report *The Transpromo Revolution: The Time is Now* by Barbara Pellow, Cary Sherburne, and Eve Pedula) describes the transpromo journey by First Data Corporation, a financial printing service bureau.

CASE STUDY:
FIRST DATA—THE OPPORTUNITIES ARE LIMITLESS

by Barbara Pellow, Cary Sherburne, and Eve Pedula
Excerpted from InfoTrends *The Transpromo Revolution: The Time is Now*
Reprinted by permission of InfoTrends

First Data Corporation prides itself on making buying and selling easier. Many people do business with First Data every day, although they may not realize it. Whether a consumer is writing a check at the gas station, using a debit card to pay for groceries, buying a book online, getting cash out of an ATM, paying for dinner with a credit card, or using a gift card to purchase something special, chances are the transaction is completed quickly and securely by First Data. In addition, First Data prints and distributes about two billion documents annually.

With its strong heritage in transaction printing and its associated data expertise, First Data is a prime candidate for the delivery of TransPromo services. Glen Wordekemper, First Data's Vice President of Product Development for Customer Communication Products, is right in the thick of this transition. He stated, "We view TransPromo as much more than just a method of cross-selling different products. It is incorporated in every interaction that you have with a customer across the life cycle of their relationship with you. This includes everything from rewards programs, loyalty, pricing, account level treatment options, and more. We have deployed a strategic communications solution (SCS) to address the TransPromo opportunity. The SCS provides us with the ability to deliver fully customized communications with unique offers that are relevant to each recipient across an array of delivery channels."

THE SALES PROCESS

Wordekemper pointed out that adding marketing information into the transaction data mix requires a hybrid business approach. While First Data executives historically worked with its counterparts in the operational areas of the financial institution, the migration toward TransPromo communications also includes key organizational experts in marketing, often at the CMO level. He stated, "Adopting TransPromo Communications requires an investment in new capabilities. As a result, the cost benefits of these investments must often be sponsored in multiple areas of the institution. Because of this, we find that there are a number of stakeholders who must endorse the overall direction when moving toward fusing transactional print and marketing concepts."

CASE STUDY

Rolling out TransPromo communications requires a paradigm shift in the way clients think about transactional documents and the potential economic value to their brand. "When we work with our customers to re-define their monthly bill or account communication, we not only focus on the clarity of the bill, but also the probable ROI which can be achieved by weaving marketing offers and other relationship-enhancing aspect into the mail piece," said Wordekemper. First Data finds that this paradigm shift is most likely to occur when marketing is actively involved in the mail piece design process from beginning to end.

Wordekemper pointed out that depending on what a customer of First Data chooses to roll out as part of its TransPromo communications, the process is likely to influence how an institution's existing marketing dollars are being spent and allocated to improve overall marketing effectiveness. Wordekemper added, "It's not necessarily a requirement to spend more on marketing, but rather a shift in the proportionate share of marketing dollars and how they are allocated from an existing marketing budget. From time to time, however, we do find that the marketing department is willing to look at additional expenditures in the context of the overall return they can generate for the business."

With a controlled test, marketers have the ability to configure test and control groups to which they can deploy different promotions. This enables them to track the success of the various test campaigns, and they can refine them based on analytics to achieve the most compelling offer possible. The most successful programs can then be rolled out on a larger scale.

TRANSPROMO IMPLEMENTATION

"What TransPromo means to First Data and its clients," related Wordekemper, "is making sure that the consumer gets the most relevant, tailored, and optimized offer, and that it is delivered in the most efficient way." Wordekemper explained that transactional data enhanced with marketing messages might be delivered via hardcopy mail, e-mail, SMS messaging, or even online self-service tools. "What is important to our customers," he added, "is that all of these media carry a consistent theme."

Wordekemper claimed that SMS as a communications medium is beginning to build in North America. He explained that one common use for SMS in an integrated communications platform is the distribution of alerts; for example, a text message may be sent to a consumer's cell phone or PDA to notify him about unusual spending patterns on his credit card or its use in an unusual geographic region. "Before," he explained, "these alerts were communicated by phone or mail, and they were often not as timely as the consumer would have liked. Now there is a two-way connection that can help eliminate fraud."

Alerts can also be used to reduce the number of customer service calls. "If you analyze the top reasons for calls, such as when the payment is due, how much is owed, and confirmation that a payment is received, you can offer consumers the ability to be reminded via alerts," Wordekemper suggested. "This ultimately is much more convenient for the consumer and saves the issuer lots of money." As consumers become more acclimated to receiving these types of SMS messages, they are likely to be more open to the addition of marketing messages to this channel as well.

Wordekemper explained that TransPromo communications occur on different levels. "It can be as easy as adding a black & white text message to a statement or notification," he stated. "Increasingly, however, our clients are turning to color to add value, to highlight different promotions, and even to gain attention and reduce cycle times from a collections perspective."

RETAIL BUYS IN

One First Data client that was an early adopter of TransPromo is involved in the retail industry.

Statements for this client base were enhanced with offers related to products the consumer had previously purchased to increase store sales. "For example," explained Wordekemper, "a consumer might have purchased a John Deere lawn tractor and could receive a $50 off coupon for fall servicing or a $100 off coupon toward the purchase of a snowblower attachment." According to Wordekemper, First Data clients are seeing significant lift from these types of targeted promotional messages included on transactional documents, but like many companies today, he declined to reveal specific results. "Suffice it to say," he added, "that our TransPromo business is growing, and it clearly works for our customers. Especially when a customer has already acted, we believe TransPromo can be a key tool to get them to act again."

First Data also offers a Web-based GUI front end that enables its customers to build their own rules to segment their portfolio, down to the individual consumer level and based on geography, purchasing behaviors, and other dynamics. As an example, Wordekemper cited a home improvement store that might include promotions on a store card or loyalty statement based on zip code, directing the recipient to the nearest store and even including a map or possibly a coupon. Once the consumer visits the store, a followup communication can be sent, thanking them for their visit and offering

CASE STUDY

10% off of their next purchase. Later, if a new store is opened in the area that is closer to the consumer, this mechanism can be used to drive traffic to the new store. This ongoing communication continues the dialogue and brings value to the consumer and the retailer.

First Data is seeing the fastest adoption of the incorporation of third party promotions occurring in the retail industry. In First Data's experience, large financial institutions are more likely to want to promote products and services from their own broad spectrum of offerings rather than including third-party offers.

TRANSPROMO: NOT NEW, BUT DIFFERENT

First Data has been conducting TransPromo in one form or another for more than ten years, particularly within the credit card space. What has changed, according to Wordekemper, is the advancements in technology that allow integration of multiple communication channels, from composition to content management and the proliferation of the Web and handset devices. "One thing that I think has flipped the switch and accelerated the paradigm shift is the do not call list," he stated. "This has prompted direct marketers to look for new ways to gain access to consumers. That has really been the igniter of the market convergence, and it is driving more interest in TransPromo."

Another technology enhancement has arisen from the configuration of First Data's Kodak Versamark printers, which include MICR capability and variable perfing. The perfing, in particular, allows greater flexibility in configuring individual statements so that one consumer might get an ad panel, the next could receive a marketing panel that offers a discount on shoes, and the next could receive an offer for a blouse. If a consumer is late with a payment, he or she may receive a payment reminder. "All of the variability and flexibility exists within our decisioning engine to design documents on the fly, and the Versamark allows us to seamlessly implement," commented Wordekemper.

In addition, with the introduction of color, coupons can present a much more appealing offer, and bar codes on those coupons enable tracking at point of sale down to the individual consumer level.

REALLOCATION OF MARKETING SPEND

Most companies are already doing some level of direct marketing, and First Data is seeing a trend for marketers to reallocate some of that direct marketing spend to TransPromo applications. Wordekemper stated, "They realize that they have a monthly appointment with the consumer, and that the consumer typically has a longer attention span when it comes to transactional documents. They might even pick one up multiple times. Our clients see a better read-and-response rate from this vehicle and are anxious to use it to obtain cost reductions and a revenue lift."

THE SHIFT TO ELECTRONIC DELIVERY

First Data's customers are also interested in accelerating consumer paper suppression. Not only does the migration to electronic-only statements reduce costs, but it offers additional flexibility with regard to statement content and format. "The electronic document can be totally different from the paper document," related Wordekemper. "Our customers have the ability to decision unique offers for the electronic channel and the paper channel. For example, on an electronic document, you would not see extra pages as in the paper statement, but you can actually include much more content by using links. We see this as an opportunity to use the data down to the individual transaction level." As an example, he cited a consumer who purchased a particular pair of shoes, prompting the retailer to market slacks or a blouse that coordinate with those shoes. "Persuading the consumer to come back and make the additional purchase is the value," he stated.

"Nevertheless, what we are really seeing in terms of the electronic communications is a one-two punch, where I send you the offer in the mail on a transaction document, but then I can follow up with an email." This process also allows marketers to test the effectiveness of the various channels and determine which channel might afford them the biggest lift for a given offer type.

WORDS OF WISDOM

Wordekemper advises organizations not to undertake TransPromo initiatives without planning for appropriate analytics and testing. "What will make TransPromo successful," he reported, "is the ability to create an offer, do test and control, get results, and analyze what made the offer effective or ineffective. It is a continuous loop of feeding the customer data into the analytics model and refining the offers on an ongoing basis. It is not a one-time thing. That is my message."

He also advises partnering with an outsource provider who has knowledge and expertise in this area to avoid the large capital investments required to do TransPromo well. "Most of all," he stated, "don't be afraid to try it. The opportunities are limitless, and there are providers like First Data who can help you exploit them."

CONCLUSION

The opportunity to personalize communications with customers, employees, shareholders, and other important constituents of businesses and organizations has given new life to many in-plant print shops. In order for these units to thrive, they must partner with their internal customers in marketing and human resource departments and work with their colleagues in the IT department to take advantage of these opportunities.

However, in-plants have the same challenge in selling these services to their internal customers as print services providers do in selling these services to their business clients. With transpromotional printing, the documents must be perfect—accurate information delivered to the correct customer or employee. If in-plants can step up to the challenge of achieving the goal of zero-defects in their work processes, they will be able to participate in this new growth area.

NOTES

140. Encyclopedia.com, "Technology of Desktop Publishing."
141. Pellow and Larsen, "It's Transformation Time for In-Plants."
142. Ibid.
143. "The Largest In-Plants," 2007.
144. Sorce, *Sourcing of Corporate Print* (PICRM-2005-02).
145. Macro, "In-Plants: The Next Generation."
146. InfoTrends, *Multi-Channel Communications: The Content Publishing Workflow Challenge.*
147. Macro, "In-Plants: The Next Generation."
148. Ibid, p. 18.
149. U.S. Postal Service, "The Household Diary Study: Mail Use and Attitudes in FY 2007."
150. Ibid, p. 27.
151. Pellow, Sherburne, and Pedula, *The TransPromo Revolution: The Time is Now.*
152. Creamer, "Think Different: Maybe the Web's Not a Place to Stick Your Ads."
153. InfoTrends, *The Future of Mail 2006: Direct Mail, Transaction, and "Transpromotional" Documents.*
154. Sorce, *Relationship Marketing Strategy* (PICRM-2002-04).
155. InfoTrends, *The Future of Mail 2006: Direct Mail, Transaction, and "Transpromotional" Documents.*

156. Ibid.
157. "PRIMIR: Trends and Future for Financial and Transactional Printing, Section VI. Consumer Research Summary,"
158. Pellow, Sherburne, and Pedula, *The TransPromo Revolution: The Time is Now.*
159. Ibid.
160. Vruno, "Print Mail's Digital Links."
161. Pellow, Sherburne, and Padula, *The TransPromo Revolution: The Time is Now.*
162. Ibid, p. 37.
163. Ibid.
164. Ibid.

CHAPTER SEVEN

TRANSFORMING PRINTERS AND PUBLISHERS INTO DIGITAL SERVICE PROVIDERS

THE DIGITIZATION OF ELECTRONIC COMMUNICATIONS has created business winners and losers. As firms such as Amazon and Google are built on these technologies and flourish in this environment, entire industries such as recorded music, camera film, and printed reference books are in a steep decline. Print is the oldest information medium and today print services providers are under siege. Since the year 2000, over 10,000 printing establishments have closed in the U.S. alone.[165] However, while some printers flounder, others are thriving. The 2007 Printing Industries of America / Graphic Arts Technical Foundation (PIA/GATF) study of profit leaders revealed that the top 25% of printers averaged a 10% profit versus 3% for other printers.[166] How should printers grow and change to meet the challenges of the digital revolution?

GROWTH STRATEGIES

There are only two ways for a business to grow by offering a new product or service: either sell the new product or service to existing customers or find new customers. The former is called *product development strategy*—the business grows by gaining new capabilities and then selling new products and services to existing customers.[167]

Successful digital color printing businesses have had a varied history of growth and development using this strategy. Many firms that originally started as commercial print providers or copy shops added digital color printing capabilities to serve their clients' needs for print-on-demand or short-run color printing.[168] Some advertising and direct marketing agencies that previously outsourced their printing needs to commercial printers now offer printing services themselves. Since the new digital color printing technology is capable of "clean" printing (minus the inks, fountain solutions, blanket wash, and volatile organic compounds that proliferate in an offset environment), these agencies can add digital production presses in-house to assist clients who want to take advantage of 1-to-1 marketing programs.

The second growth strategy is *diversification*. Using this approach, a business cultivates new capabilities to appeal to a new set of customers.[169] Start-up businesses do just that—they develop new services for new customers. Many printing firms with digital printing capabilities have used a diversification strategy by targeting new business clients who have good internal data on their own customers. Industries such as financial services/insurance firms, manufacturing firms, and retail firms have a long history of gathering data about customers and using computers to facilitate billing and buying transactions. Nearly half of the digital print services firms the Printing Industry Center surveyed for a 2004 commercial printer business model study had targeted these specific industries.[170]

Diversification can also be accomplished through an *integrative growth strategy*. Instead of developing a new capability internally, a firm might work with a supplier or a distributor to offer new capabilities.[171] For example, a print services provider that buys a mailing and fulfillment operation from a supplier is applying *vertical integration*. On the other hand, a firm that actually acquires a competitor with unique capabilities is taking advantage of *horizontal integration*. For example, an offset print provider may acquire a digital print provider, or vice versa, as in the case of the 2003 RR Donnelley and Moore Wallace merger. In horizontal integration, the buying firm acquires new capabilities by purchasing the proven capability and clients of another firm.

DIGITAL SERVICES STRATEGY

Printers can also pursue a product development growth strategy by offering ancillary services. According to a 2008 PIA/GATF report, printers' revenues from ancillary services increased from 8.1% in 2005 to 10.7% in 2007.[172] The report also stated, "Printers with the lowest proportion of total sales from ancillary services also had the lowest profit rates, and profit rates generally increase with increased ancillary sales."[173] Printers that specialized in direct mail had a higher percentage of revenue from ancillary services than all other types of printers. The most frequently offered ancillary services in the PIA/GATF study are presented in Table 7.1.

Table 7.1 *Ancillary services offered by printers*[174]

SERVICES	PERCENT THAT OFFER	PERCENT OF REVENUE
GRAPHIC DESIGN	57%	—
FULFILLMENT	54%	2.8%
MAILING	—	4.4%
DATABASE MANAGEMENT	31%	0.8%
DIGITAL PHOTO LIBRARY	17%	—
DIGITAL ASSET MGMT	16%	—
"OTHER" INCLUDING WEBPAGE PRODUCTION	13%	2.7%

Transforming a printing manufacturing business into a digital services provider business requires the commercial printer to add, at the very least, postal and email mailing services, database management services, and Web page production. Design services and photography are also great ancillary services. All of these services, including mailing, rely on the digital creation and distribution of content. Adding capabilities is difficult even if the printer pursues the vertical integration strategy by acquiring other businesses with the sought-after service offerings. This course of action is risky because of the steep investment required in personnel and new process management systems.

PUBLISHERS AS DIGITAL INFORMATION SERVICE PROVIDERS

This chapter presents four case studies of printers and publishers who have made successful transitions from being strictly print manufacturers to becoming digital services providers, either through their own efforts or through the acquisition of other businesses. We start with two cases from the world of publishing. As progressive printers have had to adapt to changing technologies, so too have their colleagues in publishing. Publishers have probably been hit harder and faster by the Internet as a competing medium for the distribution of digital information than printers.

CASE STUDY: THOMSON CORPORATION

Adding new digital services is not a quick fix, especially for firms that have a legacy of success in the manufacturing of printed products. In a recent *Harvard Business Review* article, the CEO of Thomson Corporation told of this large publishing company's success in transforming itself into an information services company over the past ten years.[175] Before the transformation, Thomson was known primarily for its printed publications—law books, textbooks, professional journals, and over 200 newspapers. As the Internet became a real threat to publishing enterprises, Thomson shifted its strategy and engaged in an acquisition and product development program to add electronic information products and software. Whereas 80% of its revenue came from printed products ten years ago, today that amount comes from digital products and services. Though it has roughly the same revenue, Thomson is now more than twice as profitable, and its revenue is, according to the *Harvard Business Review* article, "unusually repeatable, predictable, and profitable."[176]

The Thomson transformation began with the divestiture of some businesses that did not fit into the information services strategy. This in turn provided needed resources for the acquisition of businesses that did fit into this strategy. The authors of the article noted, however, that the real breakthroughs did not occur until Thomson engaged in customer research. Thomson Financial for example, sells its products to institutions such as banks, corporations, libraries, and investment firms. However, it had very little contact with the users within these firms: the employees who bought and sold stocks or provided investment advice to clients. Once this problem was identified, Thomson implemented a process for getting to know its end users better. In the first step, Thomson researchers reviewed secondary research, conducted customer interviews, and, as a result, reframed the company's segmentation. Once the new segments were defined, the researchers estimated the share of services Thomson had versus its competitors, and identified where the untapped sales potential was hiding. In step two, Thomson identified the needs of each segment and used some innovative research methodology to learn just how end users worked before, during, and after the use of a Thomson product. This research allowed Thomson to define customer workflows and the various pain points they encountered, providing insights into new products that the company could offer. For example, Thomson researchers found that "highly paid analysts were spending valuable time manually inputting Thomson Financial data into spreadsheets. So we built in a capability that allowed them to seamlessly export the information to Excel."[177]

The third step was to develop the new products that buyers (institutions) would be willing to purchase. Cross-functional teams of Thomson employees from product development, customer service, sales, and strategy departments brainstormed new ideas to test with user groups. Through this process, they further refined their segmentation strategy to develop go-to-market plans for each segment.

The implementation of this end-user research strategy developed "organically" at first, but Thomson took its successes from one division and applied it to all business units. Since 2001, Thomson has educated over half of its workforce in this key business process, and this has paid off. The *Harvard Business Review* explained:

> We estimate that nearly 70% of the products and services Thomson's businesses now offer are "nonlegacy" and have been developed through front-end strategies. In 2007 such offerings had the highest growth rates (in the double digits) and were powering the organic growth of the company.[178]

Critical to the success of the program was the involvement of the CEO as the missionary who spent time with frontline teams to develop and implement the process. In addition, the cross-functional design of the Thomson implementation teams facilitated good word-of-mouth testimonials from peers.

Thomson, one of the world's largest publishers, has had success with this transformation. Other publishers, and especially newspapers, have also had to face the challenge of becoming information services companies while maximizing their profit from the printed product. The *Democrat and Chronicle* of Rochester, New York, part of the Gannett family, is one such newspaper. The *D&C* has adapted its product to deliver news in the ways that readers want to receive it: on paper *and* on the Internet. Dr. Twyla Cummings, the Paul & Louise Miller Distinguished Professor in RIT's School of Print Media, summarizes the *D&C*'s journey below. (A video of a presentation by Michael Kane, formerly president and publisher of the newspaper, is available on RIT's Printing Industry Center Web site at http://print.rit.edu/events/miller/kane/.)

CASE STUDY: A NEWSPAPER TRANSFORMED— THE ROCHESTER *DEMOCRAT AND CHRONICLE*

by Twyla J. Cummings, Ph.D.

Prior to today's highly advanced technological age, "news" was primarily distributed via television, radio, word of mouth and of course by newsprint. The traditional newspaper business model has enjoyed extraordinary success over the last century. This business model sold space to advertisers who were willing to pay for access to readers, and sold subscriptions to readers. This represented the perfect symbiotic relationship between advertiser and reader. Advertisers had access to a known, stable audience, readers had access to inexpensive editorial content, and of course newspapers were the gatekeepers[179] However, as decreases in newspaper circulation have resulted in a sharp drop in advertising and circulation revenue, the traditional model is no longer viable. Data show that from the early 1990s to 2006, the percentage of newspapers purchased in relation to the U.S. population decreased from twenty-five to eighteen percent[180].

While electronic transmission is a major factor in the decline of newspaper subscriptions/circulation, lifestyles and generational issues cannot be ignored. As people began getting their news from other sources, advertisers began looking to other media to get their messages out to the widest audiences. Additionally, entities such as Monster.com and Craigslist have almost eliminated the need for want ads (especially for employment) and classifieds in the printed paper. Thus because of advances in technology and the shift in reader demographics, the traditional news model has changed dramatically.[181] We no longer rely on print exclusively to disseminate content.

The traditional news audience is shrinking and the structure of the companies that produce news is vastly different from what it was 25 years ago. The successful future of this industry depends on the acceptance and appreciation of new technology and methods for distributing news content.

NEW BUSINESS MODELS EMERGE

According to the American Press Institute's *Newspaper Next Report* (2006):

> Successful new business models are emerging, providing new ways to get and give information, buy and sell, create and maintain relationships, and convene communities. Many of these offerings supplant traditional newspaper functions, adding new dimensions of value,

CASE STUDY

convenience and interactivity. Consumers and advertisers are eagerly adopting these new solutions to get key jobs done in their lives. Newspaper companies that make this commitment will discover broad new horizons of civic service and business opportunity. They will succeed in engaging large new segments of customers, both among the public and among businesses, and they will discover ways to serve them more effectively than ever before.[182]

A key factor in the success of this new paradigm is leadership. If the leaders of news media companies operate with a traditional mindset new business models will not be successful (Olmstead, 2007). The *Newspaper Next* model of this shift can be seen below in Table 7.2. The information contained in the figure is indicative of how processes have changed and how information is received.

Table 7.2 *New styles for newspaper leaders*[183]

OLD WORLD	NEW WORLD
Editors select news	Consumers and aggregators select news
Communities graphically defined	Communities virtually defined
One daily source of information	Multiple daily sources of information from multiple channels
One-way advertising	Two-way system of matching buyers' wants with sellers' offerings
Consumers browse lists for information	Consumers expect precise matches to their search criteria
Advertising revenue determines success	Items sold determine success

New business models call for more diverse revenue sources. The key revenue streams for many of today's newspaper publishers are:
- **Circulation** of printed and on-line news content
- **Advertising** – both print and on-line
- **Inserts** printed by other printers and inserted into the newspaper
- **Other products** including magazines, other newspapers, inserts, and flyers
- **Other services** such as mailing and the distribution of other newspapers

Today, newspaper companies are viewed as news media companies since they no longer focus exclusively on the printed newspaper. The business model for

CASE STUDY

how news publishing companies achieve their revenues is currently in a state of flux. In a 2008 lecture given on campus at RIT, Michael G. Kane (a Gannett Company executive and formerly the president and publisher of the highly successful local newspaper) talked about the changing business model that is emerging within the news publishing industry.[184]

The *Democrat and Chronicle* (*D&C*) is the primary newspaper for the Rochester, New York area. As of March 2008, it had a daily circulation of 145,913, a Saturday circulation of 165,585, and a Sunday circulation of 199,533.[185] The parent company of the *D&C*, Gannett Co., Inc., is an international news and information company headquartered in Arlington, Virginia. According to a February 11, 2008 Scarborough Research report, the *D&C* ranked number one in

> local market...adult penetration for print audience and Integrated Newspaper Audience (defined as the weekly net unduplicated audience of the printed newspaper and its website.) Seventy-nine percent of adults in the Rochester, NY local market read the printed version of the [newspaper] during the past week. Together with its website, www.democratandchronicle.com, the publication reaches 81% of the Rochester market on a weekly basis.[186]

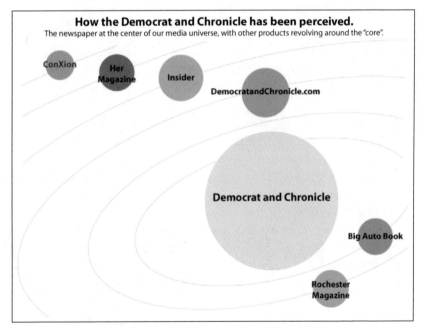

Figure 7.1 *Transitional business model for the Democrat & Chronicle*[187]

158 PERSONALIZATION

CASE STUDY

The current business model for many newspapers is similar to the one illustrated in Figure 7.1. This model depicts an earlier business strategy of the D&C where the core product, the printed newspaper, was the central focus with a few other little products revolving around this primary revenue source. While this was a shift from the traditional model, it did not yield the successes and the revenues needed to stay competitive with other news sources and did not effectively keep non-core audiences engaged.

Newspaper publishers realize that they can't be married to the "newspaper" if they are going to remain viable businesses. Successful companies are beginning to look very different from the transitional model discussed earlier. To quote Michael Kane, former president and publisher of the Democrat and Chronicle "newspapers are still labeled by SIC (Standard Industrial Classification) as a manufacturing sector and that needs to change... While print is not dead it's growing in different ways."[188]

A successful and appropriate business model for newspapers involves a diverse mix of product offerings. The Democrat and Chronicle has taken this approach which is illustrated in Figure 7.2. To accommodate this business strategy, the newsroom at the D&C has been restructured to that of an information center. The center operates 24/7 and the focus is now on inte-

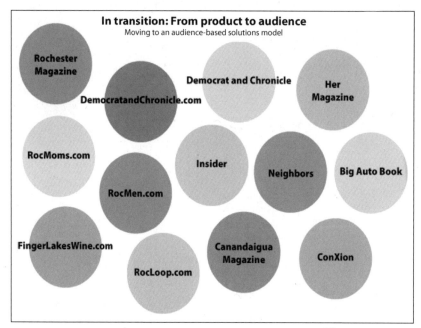

Figure 7.2 *Democrat & Chronicle's transition from product to audience*[189]

Transforming Printers and Publishers into Digital Service Providers

CASE STUDY

gration and the use of all media platforms to disseminate content.

A review of the 2008 Newspaper Next report further substantiates the need for newspapers to diversify in a fashion similar to the one shown in Figure 7.2. According to the report, today's newspapers are working very hard to create new products and revenue streams. The Newspaper Next study describes tools and processes developed by the American Press Institute where the vast majority of these innovations are close to the core and lean entirely on traditional revenue models. However, the goal is to attract new target audiences. The report emphasizes that the innovations must take place quicker in order to offset the rapid decline in traditional revenue streams[190].

Newspapers are no longer thinking of just subscribers, but audiences

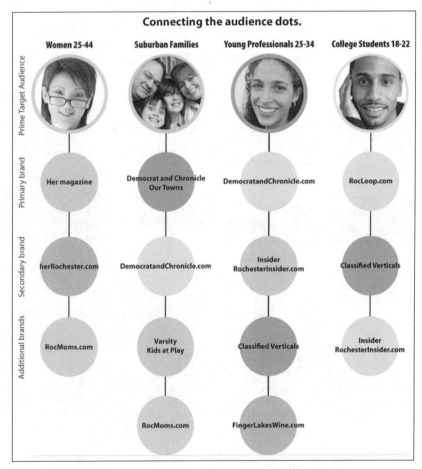

Figure 7.3 *Democrat & Chronicle's in-depth focus on audience*[192]

and are working to produce products for a cross section or wide variety of audiences. Figure 7.3 illustrates the D&C's approach to this concept. According to Kane there is a huge cultural component to be considered relative to audience. "To aggregate audience you have to follow the audience.[191]"

GOING FORWARD: THE FUTURE OF THE D&C

Most industry reports suggest that the future outlook for the printed newspaper is bleak. This is largely due to the impact of the Internet and other electronic media as an alternative. Additionally, as the age demographic changes and the core readership declines, there are a growing number of young people who are not interested in news or prefer to get it elsewhere. Given this outlook it is imperative for newspaper publishers to diversify their business models and incorporate a strategy that is consistent with the way people live and with how technology works today.

The Rochester Democrat and Chronicle has risen to the challenge of new media and has changed the way in which it distributes news content. Going forward, the company realizes that more changes will be required as technology options increase and demographics continues to change. While D&C executives realize that "print is not dead" and they will not ignore the core product, they are also not willing to rest on current successes. The future audiences and technologies will dictate their product offerings and target audience. According to Michael Kane, this future includes[193]:

- smaller newspapers
- print and digital convergence
- strategic partnerships and acquisitions
- partnerships with schools
- embracing social media networks

Kane also states that despite the physical changes in the paper and enhanced product offerings, we should not forget "that even though the nature of journalism will change original reporting is still important." [194]

PRINTERS AS DIGITAL INFORMATION SERVICE PROVIDERS

Publishers have control over content creation, which they can monetize through restricted digital distribution. Their long history of content ownership may provide them with an easier transition to the digital services era where content is king. Commercial printers, on the other hand, are most often the outsourced manufacturers and distributors of content provided by clients. While this content is often for the purpose of marketing, some printers have had a history of providing printing for other business needs. In this section, we'll provide a case history of one small commercial printer that specializes in marketing applications (Global Printing, Inc.) and a large printer that specializes in business process improvement applications (Standard Register). Both companies are transforming their businesses to become digital services providers.

CASE STUDY:
GLOBAL PRINTING, INC.

In the summer of 2008, Jon Budington, CEO of Global Printing, Inc. in Alexandria, Virginia, had a problem. "We are on the constant quest for people," he said. "Global grew by 27% last year (2007) and 15% the previous year."[195] Budington created this growth by opening a fulfillment center and a digital printing operation, by offering Web design and data management, and by acquiring a competing printer with a successful a mailing operation. How has this small firm made such a big leap to providing its clients with a single source for integrated print and online communications?

HISTORY

Global Printing, Inc. was established in 1979 by Jery Dreo. The bulk of the company's early business was short-run, black-and-white paper printing. The year 1999 was the high point for Global Printing as a traditional print provider. At that time the company's strategy was to focus on its equipment list and to solicit bids. With the explosive growth of high-tech firms in the Washington, DC area, printing jobs were plenty and profitable.

Jon Budington joined the company in 1991, upon graduating from RIT's School of Printing. He rose rapidly through the ranks, holding five different positions in the company with increasing responsibility. In the aftermath of 9/11, Budington was Global's director of operations, responsible for keeping the bank at bay as credit got tighter and revenue dropped by approximately one-third, from over $6 million to just under $4 million. In 2002, he was promoted to CEO and began making radical changes to the company.

While many other printers were cutting prices to keep the presses running, Budington decided instead to change Global's business model. First, he cut jobs. For example, by modernizing the accounting software, moving the billing responsibility to a customer service employee, and eliminating all unnecessary data collection, he cut the accounting and estimating staff from seven to two people. Once he felt that he had attained the right number of employees for the business (a staff of 35–40 people), Budington ventured onto a new path: he became the leading sales rep. His first job was to conduct customer audits on all of Global's customers in the second quarter of 2002. His goal was to determine what percentage of their print budgets Global currently provided, and then, for each client, to increase the "share of wallet" for printing that went to Global. He needed to grow Global's top line revenue, and selling more to current customers was his focus.

CASE STUDY

As Budington talked with his clients, he realized that printing was just one part of the value chain of customer communications that they were managing. For example, when Budington was selling print collateral to one client, he asked questions about the challenges that client faced to keep his collateral in stock and up-to-date. That immediately moved the dialogue to a print-on-demand solution that Global could deliver in addition to the printed product. A periodic $10,000 job from this client changed to a much higher annual revenue stream.

Budington developed a business strategy that embraced the firm's manufacturing heritage but moved beyond printing to provide additional services. First, he bought Serif Press, a small printer with a large mailing operation and one large client (Access Intelligence). Realizing that the Internet was changing the way customers and businesses used information, Budington embraced it and expanded his business to offer Internet services like email and data management. For example, when he acquired the Motley Fool publishing account, Global printed and mailed the newsletter "Hidden Gems" for its subscribers. As the subscriber list grew, so did the pile of "undeliverable" returned mail at Motley Fool headquarters. Budington offered to help with that headache and changed the return mail address to his own. As the mail now piled up at Global, he began adding a 2D barcode to the address field during the mailing process. (2D barcodes contain more information than conventional, "one-dimensional" linear barcodes.) Any returned mail was quickly scanned and sent to Motley Fool to deal with the incorrect address issue. Global then went a step further and built an email application that could notify subscribers of their incomplete or incorrect mailing information. The goal was to increase client satisfaction for Motley Fool's subscribers. Global also utilized the U.S. Postal Service's National Customer Support Center (NCOA) database to update addresses.

But this is not where it ended. Global now helps the publisher cross-sell other product offerings to its subscribers through a multichannel communications program. The program begins with an email to one-quarter of its subscribers, informing them of new books on investing. If a subscriber does not open the email after three "blasts," Global sends a printed postcard with a URL announcing the Web site. Subscribers that still don't check in get a larger, more richly printed mailing. Global now manages the complete client contact process, and Budington has a place at the table when Motley Fool plans its annual marketing strategy.

CASE STUDY

WALKING THE TALK

In order for Global to make this transformation, it had to practice what it preached. One of the first operations to change was client interaction through the sales force. Budington hired a vice president of marketing who used Salesforce.com to collect data on Global's clients. Global draws on this data to deliver custom print marketing collateral and email communications to its clients. While many printing companies see sales and marketing as a single effort, Budington uses his marketing department to collect information and create sales messages, and his sales force to deliver these messages. By utilizing Planet Press, Fusion Pro, and some custom-written software, Global produces data-driven communications for its clients and its own marketing needs.

With the successful development of these "platform agnostic" communications tools, Global now needed to get involved in the creative process, since, as Budington explained, "many of our non-conventional ideas require our being involved with the design of the message vehicle." To do this, he convinced a long-term creative client to move its office into the Global offices. Accelerant Studios is now a partner in the sales process, giving Global the ability to develop the strategy, design the device, manufacture the device and distribute the message.

INVESTING IN PEOPLE

The work has also changed for the people employed at Global. In 1999, Global had 12 people in sales and 4 in customer service (a 3:1 ratio of sales reps to support personnel). Now the ratio is reversed; Global has 9 people selling (including Budington) and 18 others who provide customer service and IT support.

One example of employee transformation that supported the organization's change is Kevin Fay, who joined Global in 2004 after he completed his B.S. degree in Printing Management from RIT. Fay began as a special projects coordinator who was responsible for Global's "crazier" jobs. These included projects that needed special finishing, fulfillment, or database work.

When Fay began cleaning and formatting client files to import into the mailing software, he encountered problems such as incomplete records, hard returns, weird characters (i.e., asterisks) in data fields, and no spaces in address fields between street and city. Fay put standard processes into place, and now data cleansing and mail preparation work is being completed by the new mailroom supervisor. When Fay noticed that clients were very casual about sending databases via unsecured email attachments through the Internet (some of these files even contained social security numbers), he set up an encrypted, password-protected SFTP server to which clients can upload their files.

Transforming Printers and Publishers into Digital Service Providers

CASE STUDY

Next, Fay learned MySQL and JavaScript, leaning heavily on the PHP scripting language class he had taken in college. He put this to good use in building an elementary online ordering system for a steady client so that the client could order items against inventory. Fay built the basic graphical user interface (GUI) for the client's sales customer relationship management (CRM) program using JavaScript, with MySQL as a database backend.

With Global's purchase of a Kodak Nexpress, Fay took advantage of the one week of free training on the variable data Planet Press software and the RIP. He immediately put this to use in creating an innovative solution for the HR department of one of Global's clients. That client was losing trained employees to competitors who were paying higher salaries but had non-competitive benefits packages. The HR department wanted to communicate with its existing workforce the extent of the benefits that each person received that were actually greater than an attractive salary increase. Global designed a total compensation statement, a four-page document customized for each employee that presented personalized data in charts and tables with dynamic text. This was a great success for the client firm, which found that employee turnover decreased after the program was started.

The number of jobs at Global requiring Internet and database skills continued to increase, and in 2008 Budington hired a new IT manager. Global also purchased Printable, a Web-to-print system, to replace its homegrown system. Printable can be used for more clients and can secure work via client login and passwords. Using its new capabilities, Global sold an innovative variable data postcard campaign to a local bank client. With it, the bank mails customized postcards to bank customers, letting them know where the closest ATMs are to their homes. These new applications are possible because of the additional capabilities that came with new employees and the continuing development of current employees.

THE NEXT STEP

The next step in Global's transformation was a name change. In spring 2008, Budington bought the domain name "Global Thinking" to serve as the name of the new business model that had emerged at Global Printing. Along with the name came two new employees. Jason Kowal as the new chief marketing officer, provides marketing research and database management services to Global's clients. "My job at Global is to design a process for interactivity between a client firm and its customers," Kowal said. He believes that most mid-sized companies do not know how to use personalized communication. Their customer lists are limited and, while they may provide digital printing, they don't realize that with digital comes the opportunity to sell personalization and dialogue marketing using email and the Web. Kowal noted that one client had a 50% bounce-back rate for its email and postal mailings. When a client has bad data, Global steps in to do the very unglamorous but necessary job of data hygiene.

This consultative sales process is also longer than the typical print sales cycle. As Kowal explained, "You have to provide the ROI on VDP. We must make the case by providing benchmarks from our historical successes with other clients." If a client has a good database, such as those that are found in human resource and medical benefits service providers, it might take 30 days to close a sale. For marketing communications services, it will take 90-120 days to close because the first step is data integration and enhancement of the client's database. The clients must see Global as a trusted partner in their customer contact strategies, and this takes time and early successes.

Often, the first part of a program requires Global to obtain information directly from the client's customers to enhance the database. As Budington says, if you ask for personal information, you must give something back: "Quid pro quo." Global uses a printed piece to drive the client's customers to a Web site where they complete a registration or survey form. If they do this, they each get a gift, which takes advantage of Global's fulfillment operation. The early success of this strategy (as shown by the number of people who complete the forms) is just the thing to demonstrate Global's skill and delivery on its marketing programs. This sets the stage for the next part of Global's service delivery, the result of which develops an annuity program with a client.

A key player in this consultative sales process is Global's new account executive, Shari Fox, who had over 20 years of print sales experience before she came to Global. What she is excited about is being able to bring Global's mix of services to her long-standing clients. For example, a local private university asked Fox to bid on a print job for the admissions process. She and Budington called on officials at the school to respond to the bid, but also to better understand what the printed product would be expected to do. They found out that this university's applications had dropped by over 25%. While the conversation started with printing, it evolved into a discussion of how to increase applications and target international students. Global is now working on a complete communications proposal for the university that leverages both its Internet and print expertise.

CASE STUDY

CHALLENGES

Fast growth brings a number of challenges. First and foremost is that Global has outgrown its management structure. As more and more work comes in, support personnel, print production managers, and fulfillment operations are pressed to their limits to juggle old and new work, often on tight deadlines. As Budington looks to grow his employee base, he is back to the challenge stated at the opening of this case study. Using the analogy of a muffler retailer who evolves to the point of offering full care repair services, Budington says, "We are looking for total car service people, not just muffler shop guys." His management challenge is to keep people who have performed well in traditional print work and are now working for new middle managers with the broader scope of "Global Thinking." It will be a difficult process to keep veteran employees motivated as they see their hopes for career trajectories changed with the new hires. Creating new career trajectories will be required to show veterans how they can contribute and grow in the new business model. Internal customer service employees must be given opportunities to participate in direct contact with clients to build relationships so that they too can develop new service offerings to be delivered by Global. That will take mentoring by senior management, whose time is scarce since they are doing all of the selling.

Another challenge is the competition. While most printers do not offer the full range of services that Global does, they can be cutthroat with pricing their printed products. Budington responds to these underbids by offering some printing for free for the clients he is serving with annual campaign and data management. As Global increases its scope, it is also meeting new competitors, such as advertising agencies. It is not unusual for Global to be the only "printer" in the room during an RFP (request for proposal) briefing. This brings a new challenge to Global—does the company want to be a full media services agency and handle broadcast advertising?

Budington delights in these challenges. As he said at the close of our interview, "I'm having the most fun, making the most money and doing the most interesting work. I never want to go back to the way we worked ten years ago."

STANDARD REGISTER MOVES INTO WEB-TO-PRINT

John Q. Sherman founded the Standard Register Company in 1912 in Dayton, Ohio, to bring to market the autographic register, which became the standard for paper feeding. Standard Register showed its competence in process innovation when it unveiled Paper Simplification, a formula to help customers improve their business processes. Printing business forms was the heart of Standard Register's business.

The organization went public in 1956 on the NASDAQ, and in 1996 Standard Register stock began trading on the NYSE under the symbol "SR." In 1983, Standard Register entered the print-on-demand arena with its first "Stanfast" center. Signifying a significant change in strategy, the organization changed its focus to being a "strategic thought leader" in 2002, focusing on various consulting services to adapt to the changing times in the printing industry. An example of one of the company's newer Web-to-print services is provided below.

CASE STUDY:
STANDARD REGISTER/DEALER OFFICE XPRESS (DOX)

In recent years, new variable data applications and the Internet have enabled a printer's customers to easily create targeted marketing campaigns. However, developing a system to facilitate this capability can be a major undertaking, especially for a printer without an extensive IT staff. To assist printers, companies are developing the necessary applications and then licensing their use to printers and their customers. This is commonly referred to as an application service provider (ASP) model.

We recently had the opportunity to talk with Chuck Avery of Standard Register's Dealer Office Xpress® (DOX) group to see how Standard Register is helping auto dealers more easily create personalized marketing communications over the Internet through its ASP system, DesignOnDemand® (DOD). Standard Register uses its DesignOnDemand application to support automotive, truck and powersports dealers with their targeted marketing efforts. Dealer Office Xpress is an alliance between ADP and Standard Register, two leaders in automotive and document solutions.

Currently there are more than 20,770 car dealerships in the U.S., most with information systems in place to manage their day-to-day business. Approximately 8,500 of these dealerships use ADP Dealer Management Systems. The solutions ADP offers are gaining popularity.

The capabilities of ADP's Dealer Management Systems mirror those of an enterprise resource planning (ERP) system, including modules to manage sales, vehicle financing, service, and accounting, as well as a CRM tool to support marketing and campaign management.

Through the DOX program, Standard Register and ADP are key suppliers of business forms, office supplies and marketing tools to automotive, truck and powersport dealers. ADP's w.e.b.CRM application provides a Campaign Management tool which is partnered closely with DesignOnDemand product, through DOX. Powered by Standard Register's SMARTworks® technology, this powerful Web-based marketing tool was already being used in a variety of industries including insurance, healthcare, manufacturing and financial services to automate production of highly personalized marketing collateral.

Integrating ADP's w.e.b.CRM tool with DesignOnDemand®, through DOX, applications enables dealerships to take advantage of their customer information to create customized customer communications. The system provides a beginning to end fulfillment solution that delivers full, four-color communications and promotions within 72 hours of order placement.

CASE STUDY

In 2003, Standard Register and ADP created the variable data printing DOX solution, using customer relationship management (CRM) and other dealer management systems. The programs goals are:
- to provide dealers with the ability to build lasting customer relationships by bringing customers back into the showroom and service departments,
- to allow dealers to harness the power of their databases to specifically target individual customers or prospects,
- to communicate special offers quickly and more efficiently, from thought-to-execution-to-print-to-distribution within 72 hours, and
- to simplify marketing campaign development and execution with a system that dealers can use right at their desktops.

DesignOnDemand allows dealers to create 1-to-1 print communications. During the development of the system, a key task was to create templates for personalized documents such as postcards, letters, brochures and coupons, and to acquire digital assets such as logos and photographs of car models. DesignOnDemand, through DOX, currently has 23 design templates for such products as jumbo postcards, small postcards, letters, and bi-fold self mailers, and gives dealers access to over 6,000 car photographs.

Instead of incurring the expense of gathering photographs of all models of all new cars in all the available colors, Standard Register turned to AutoData, a company with a complete library of automobile images. A license agreement with AutoData makes it possible for dealerships to choose from any of the available pictures at the time they need them

DesignOnDemand is useful to various dealership personnel, including parts & service managers, marketing managers and sales managers. To get started, a dealership must have access to the DOX SMARTworks®, through ADP's DealerSuite.com. For this access, DOX charges a one-time installation fee and a monthly support fee for application maintenance. DesignOnDemand allows dealers to control their own messaging and branding, reducing advertising agency costs and the development, print and distribution cycle time that agencies typically require. See Figure 7.4.

Dealers can query their own databases to find the exact population they want to target (e.g., those who need an oil change). Then they simply select the desired design template (a reminder postcard), customize it, and place the print order. A high resolution PDF file is transmitted to Standard Register and is automatically queued for printing.

Another timesaving advantage of the system is that it eliminates the

CASE STUDY

need for securing approval from senior management or corporate offices for every marketing piece because the design templates already integrate client-approved messaging and branding. Users can quickly create and send their mailings without delays.

Dealers can use the system to send customers sales promotions, new vehicle announcements, and service reminders for their vehicles. Although vehicle service accounts for only 12% of a dealership's revenue, this is actually represents nearly 50% of its profit, so targeted mailings can provide dealers with an excellent return on their investment. DOD's targeted approach to marketing is attractive since car dealerships are facing increased competition from "quick lube shops" and small vehicle repair companies. Dealers can also alert customers when their leases are about to expire, and offer them incentives for signing new leasing agreements.

Once a DesignOnDemand order is submitted, it can be routed to one of Standard Register's five print centers in Connecticut, California, Illinois, North Carolina or Texas. A variety of digital printing engines, including the Xerox 2060 and iGen3, and Kodak NexPress 2100 then print the materials. The prices for these applications are slightly higher than a typical mail-merge variable data application, but since most dealerships use higher levels of personalization anyway, the extra cost is justified by the ROI of the campaign.

Currently, DOX has 1000 dealers signed up to use DOD software, and several hundred more are in line to install it. Standard Register has also customized DesignOnDemand to work with other dealer management providers, and more enhancements to the DOD system itself are on the horizon.

CASE STUDY

Figure 7.4 *DOX variable data mailing piece*

CONCLUSION

This chapter reviewed four very different firms, all with a legacy of manufacturing print-on-paper products, that transformed their businesses by adding new digital services to their product mix. In every case, the transformation started with the owners and managers of these businesses, who made the commitment to change and then saw their businesses through the complete process, often functioning as missionaries for their vision. The process then turned to understanding the customer. These firms studied the ways their customers used their printed products to see how they could take advantage of digital distribution to save their clients time and money, or to deliver a better method of presenting and using information within printed documents. Lastly, these firms invested in their people to give them the skills they needed to create new services. These changes are not radical; rather, they are necessary steps for business transformation.

NOTES

165. Davis, "Update on Printing Industry Restructuring: Long-and-Short-Term Trends and Their Competitive Implications."
166. "2007 PIA/GATF Ratios Show Printing Industry Profits Increasing."
167. Kotler, *Marketing Management*.
168. Sorce and Pletka, *Digital Printing Success Models: Validation Study, 2004*.
169. Kotler, *Marketing Management*. Two other growth strategies also involve selling more existing products. One is *market penetration*, in which a company sells more existing products, or more services, to existing customers. The other is *market development*, in which a company sells existing products and services to new customers, often by expanding geographically.
170. Sorce and Pletka, *Digital Printing Success Models: Validation Study, 2004* (PICRM-2004-06).
171. Kotler, *Marketing Management*.
172. Davis and Gleeson, *Expanding the Market Space: Printers' Diversification into Ancillary Services*.
173. Ibid, p. 12.
174. Ibid.
175. Harrington and Tjan, "Transforming Strategy One Customer at a Time"
176. Ibid, p. 64.

177. Ibid, p. 67.
178. Ibid, p. 72.
179. Cummings, Vogl, Casonva, and Borlado, *An examination of Business and Workflow Models for U.S. Newspapers.*
180. Newspaper Association of America, *"Total Paid Newspaper Circulation, 1940-2006."*
181. Cummings, Vogl, Casonva, and Borlado, *An examination of Business and Workflow Models for U.S. Newspapers.*
182. American Press Institute, *"Newspaper Next: Blueprint for Transformation."*
183. Olmstead, "New styles for Newspaper Leaders: Do You have What It Takes to Lead Change?"
184. Kane, "Sprinting a Marathon: The Newspaper Transformation Today."
185. Audit Bureau of Circulations, "Print Circulation Trends."
186. Reuters, "Scarborough Research Releases Newspaper Penetration Report."
187. Kane, *"Sprinting a Marathon: The Newspaper Tansformation Today."*
188. Ibid.
189. ibid
190. American Press Institute, "Newspaper Next: Blueprint for Transformation."
191. Kane, *"Sprinting a Marathon: The Newspaper Tansformation Today."*
192. Ibid.
193. Ibid.
194. Ibid.
195. Budington, Personal interview, September, 2008.

CHAPTER EIGHT

MEASURING SUCCESS: CLOSING THE FEEDBACK LOOP

IN ORDER FOR A COMPANY TO COMMIT to the level of investment required for implementing personalized marketing communications, top management must first understand the value of these programs. In order to prove the value, outcome measurement is required: evaluating how personalized advertising compares to standard marketing communications regarding revenues and profits. Measuring the impact of advertising has been a top concern of advertisers for over 100 years. A now-famous quote by John Wanamaker, a department store owner in the late nineteenth and early twentieth centuries, demonstrates this point: "Half the money I spend on advertising is wasted; the trouble is I don't know which half." [196]

A large part of the challenge is that advertising is only one part of the marketing function that influences consumer buying. A firm vies for customers in an intensely competitive marketplace where price, distribution, and product differences all work together to differentiate the offering to consumers. Advertising is only one of many promotional tools that marketing managers use to provide information and incentives to target markets. The other tools include personal selling with supported collateral material, promotional incentives to resellers, and public relations activities. Many large firms rely on advertising agencies to provide an overall promotional strategy, as well as to develop creative elements, recommend media, and implement the approved plan by producing the advertising and then buying space or time in the media selected. Isolating the impact of advertising or a single campaign amidst this complexity is very challenging.

A MYRIAD OF METHODS

Measuring the effectiveness of a marketing program is a challenge that has been with business for decades. The PIMS (Profit Impact of Market Strategy) business information database, available at www.pimsonline.com, gives the real-world business performance experiences of more than 3,000 businesses, representing 16,000+ years of data.[197] A quick look at the wealth of this PIMS data will reinforce the fact that this measurement issue is not a simple one.

Assessing the impact of a marketing program can be carried out at the boardroom level (by asking, for example, whether marketing expenditures can be linked to overall financial measures such as shareholder value), or at the tactical level (again by asking, for example, which changes in a marketing promotion produced the revenue achieved). The measures assessed can be long-term in scope, such as brand asset valuation, or short-term, such as tabulating the number of coupons redeemed for a specific promotion. The measures can be tracked internally (i.e., tracking the number of hits on a Web site), or outsourced to a firm such as a trade association that tracks total market size across all competitors in an industry in order to determine an individual firm's market size. The measures can be simple counts, such as how many prospects return a reply card, or they can be the result of complex statistical modeling to predict sales volume as a function of marketing expenditures, competitive responses, and overall economic factors.

In general, most large corporations use more than one marketing metric.[198] In a study of over 700 of the leading advertisers in their industries, approximately four metrics were used per corporation in annual reports to respective corporate boards. The most frequently used metrics were:

- market share (79%),
- perceived product or service quality (77%),
- customer retention (64%), and
- customer profitability (64%).

A different survey of a small number of leading U.S. firms concluded that the most popular metrics of overall marketing performance at the strategic business unit (SBU) level were market share, sales, and profits.[199] The choice of metrics should be determined by the nature of the firm, the competitive strategy, and the particular marketing challenges the firm faces.[200]

Once a measure or set of measures have been determined, there remains an additional challenge: *predictive validity*. Can we associate the outcome measure (market share, revenue, or profits) to a change in the marketing strategy (for example, increased spending on advertising)? How can we separate the impact of an adjustment to advertising spending from other factors in the environment, such as improved economic conditions or a change in a competitor's marketing strategy? Complex econometric models have been designed to quantify the contributions of marketing mix factors to the financial performance of a firm.[201]

For marketers without statistical expertise or solid database resources, the ability to use complex modeling is limited. Many marketers use simpler metrics to assess the impact of specific marketing communication tactics on measurable outcomes within a limited time frame. For example, the impact of a local Super Bowl promotion (*specific tactic*) can be calculated by noting the change in weekly sales (*outcome measure*) before and after the promotion has run (*time frame*). Common tools used to gauge the effectiveness of advertising and other promotional tactics are outlined in the following section.

ADVERTISING EFFECTIVENESS MEASURES

Accountability is the watchword for marketing and advertising managers in today's business environment.[202] This comes after years of relatively uninhibited expenditures on mass printed and broadcast advertising, where the effects of the advertising are difficult to link to specific purchasing behavior. Today, a marketing manager will have a hard time demonstrating conclusively that investments made in marketing (and, specifically, in advertising and promotion), have improved outcome measures such as revenues and market share. Though direct marketing has historically been a strong marketing tactic, strict accountability is now being asked of it and all other marketing efforts.

Assessing the impact of a specific element of the marketing mix (for example, a certain campaign, or a price or packaging change) has become easier with the widespread use of *point-of-sale* information tracking, which provides a good outcome measure for products that are low in price and purchased frequently. For more expensive items that are purchased infrequently, a sales measure taken at a particular time may not be accurate because a lot of road is covered between a buyer's first exposure to a particular advertising message and his or her eventual trip to the cash register. Buying is the end-point of a marketing communications process that is designed to create awareness, establish a preference, and then motivate the customer to buy the advertised product or service.[203]

One theory of how advertising impacts a media user is called the *hierarchy of effects*. This model starts with advertising *exposure*, a measure of the target audience's opportunity to see the ad while consuming media. This stage is under control of the advertiser, who determines the message, and then selects the media and amount of market coverage (a major determinant of the media budget). From this point forward, the media users take control by determining what information they attend to. The audience's *attention* to advertising varies by differences in interests and need states, in a process known as *selective perception*. If the information is relevant, the media user's focal attention is captured and the information is encoded into his or her active memory. Once the ad is encoded, the message content within the advertisement is represented by a change in the media user's mental state as reflected in message *comprehension*. Comprehension can be classified as *cognitive* (a change in awareness, beliefs, or knowledge), *affective* (emotional or attitudinal variables such as liking, preference or trust), or *experiential* (through interactions with the product itself). The experiential factor is particularly effective when people are presented with the trial use of products that can be part of the promotion mix. In other cases, experience is built through interactions with another's consumption of the product (i.e., riding in a friend's new Lexus).

Other behaviors that precede the final stage of buying the product are inquiry and "request for more information," both of which are often identified as goals of business-to-business (B2B) marketing. A consumer's mental image of the brand certainly influences the likelihood of purchasing the product. Two common metrics of the actual purchasing stage are the first-time purchase and the repeat purchase. Though the hierarchy of effects model has been disputed recently as a reliable method for measuring the effectiveness of advertising,[204] it still is a useful starting point when articulating the communication goal of advertising during the planning stages of a marketing campaign.[205]

Well-designed promotional programs always identify the communication outcome in the planning stages of a campaign. These measures are determined not only by the communications objectives of the campaign, but also by the nature of the product and its current stage in its *product life cycle*. For new brands or products, awareness is a common communication objective. For existing brands or products with longer purchase cycles, changing consumers' preferences is the objective. For products that are purchased frequently, a change in sales may be the stated objective of the marketing campaign. Media metrics and advertising accountability measures can be classified by what stage in the hierarchy of effects the ad message they intend to influence. The following section discusses common advertising effectiveness measures.

Media Metrics and Advertising Impact Measures

Media metrics have one of two broad purposes: 1) to sell media time or space to advertisers, and 2) to determine the impact of the advertising. The metrics used to sell advertising can be collected by the broadcaster or publisher that monitors its own audience, or through the use of syndicated research services that quantify and determine the nature of the "tuned-in" audience. For example, the Nielsen organization has measured TV viewership for decades. TV broadcast networks use these data to describe the percentage of viewers who have tuned into a certain TV program and therefore have the opportunity to see advertisements broadcast during the program. The larger the audience and the more attractive the segment tuning in to a program, the more easily the broadcaster will be able to sell advertising within this program.

Mainstream media metrics have focused almost exclusively on advertising exposure in the effort to sell media time and space: How many people in a specified demographic audience have the opportunity to view the ad? This measure indicates the media's *reach* and is crucial for the initial selection of the media in an ad campaign. A list of many of the major syndicated services for each medium is given in Table 8.1. Typical methods for capturing these data include circulation figures and diary methods, where a sample of households report what media they are watching during a given time period.

The metrics used to assess advertising impact can determine whether a sample of people have paid attention to an ad, comprehended its meaning or engaged in some type of behavior after viewing or hearing it. As media users ourselves, we know that there is a huge gap between media exposure and paying attention to an advertisement. For example, the number of people who receive or buy a newspaper on any given day is much greater than the number who read a specific advertisement in a specific section. By the same logic, the number of households that have a certain TV program turned on is usually larger than the number of household members present during the airing of a particular commercial message within that program. At the same time, the widespread adoption of digital video recorders and the easy availability of video and DVD movies have reduced the amount of broadcast television that households watch.

In order to measure the impact of advertising in mainstream media, primary research must be conducted with a target audience. Individuals who have had an opportunity to view or hear an advertisement must be contacted directly to ascertain the degree of their recollection of the ad. There are a number of syndicated services that do this, and Table 8.2 presents the major vendors.

Table 8.1 Media metrics and related syndicated services

MEDIUM	TYPICAL MEASURE	SYNDICATED SERVICES OR IN-HOUSE METRICS	COMMENTS / NOTES
RADIO	Quarter-hour exposure to radio broadcasts, computed four times a year.	Arbitron's RADAR (Radio's All Dimension Audience Research)	Arbitron uses two methods for obtaining data: 1.) Self-administered daily diaries; 2.) Portable People Meters (PPM), which electronically detect encoded tones in broadcast media. Users wear them all day and 'dock' them at night to recharge and transmit data back to Arbitron.
TELEVISION	Nielsen ratings (the size of audience, expressed two ways: as a household rating and as the number of individuals watching)	Nielsen Media Research	Interpretation: A rating of 7.9/12.7 means that 7.9% of all U.S. television households were tuned in to the program; and this represents 12.7 million viewers over the age of two. As of September 1, 2008, there were an estimated 114,500,000 television households in the U.S.
LOCAL NEWSPAPER ADS	Response rates, circulation	Scarborough Research; Audit Bureau of Circulations (ABC); Research and Analysis of Media (RAM)	RAM measures ad exposure through reader panels and surveys
MAGAZINES	Circulation	Audit Bureau of Circulations (ABC); in-house tabulations	Publications know their own circulation, ABC verifies circulation information
OUTDOOR (BILL-BOARDS)	Number of vehicles passing per hour, day, month, etc.	State or local department of transportation (DOT); Traffic Audit Bureau for Media Measurement Inc. (TAB)	TAB measures the number of people who are exposed to out-of-home signage. This calculation, known as Daily Effective Circulation (DEC), matches billboard inventory submitted by plant operators against traffic counts published by local DOTs.
WEB SITES	Hits, unique visits	Google Analytics; Web hosting service (ISP); comScore, Inc.; Nielsen Media Research (Nielsen/NetRatings or NNRs)	User panels are used by both comScore, Inc. and Nielsen. Some provide more detailed information on who visits a site (age, income, sex, ethnicity, etc.), in addition to number of visits.
EMAIL	Responses	In-house tabulations	
CROSS-MEDIA	Surveys	MediaMark Research & Intelligence	Looks at a host of media-related activities. The *Survey of the American Consumer* presents data from 26,000 face-to-face, at-home annual interviews.

Table 8.2 *Typical advertising impact metrics and syndicated services by medium*

MEDIUM	TYPICAL MEASURE	SYNDICATED SERVICES	COMMENTS / NOTES
RADIO	15-minute time slots, PPMs, computed quarterly	Arbitron	PPMs monitor encoded inaudible signals; quarter-hour listening data obtained quarterly through phone/mail diaries (see Table 8.1)
TELEVISION	Viewer ad recall; viewing hours; channels/programs watched	Burke, Inc., Nielsen Media Research,	Nielsen tracks commercials/viewing habits by a diary; Burke's Day-After Recall (DAR) test is conducted by telephone
MAGAZINE	Three measures: 1. noted (saw) the ad, 2. brand association, 3. read most (half) of it	Roper Starch Worldwide	Realistic research setting where people read magazines and then are queried about each ad
INTERNET	Action (download, registration, clicks, etc.); keyword searches	comScore, Inc., Nielsen Media Research, Google Analytics	comScore, Inc. and Nielsen use panels to obtain more complete data than Google.
	Visits/page views	comScore, Inc., Nielsen Media Research, In-house tabulations, Google Analytics	While Google Analytics, in-house tabulations and any hosting service can provide these statistics, comScore and Nielsen also give the profiles of respondents (demographics, etc.)
	Time spent	comScore, Inc., Nielsen Media Research, In-house tabulations, Google Analytics	
EMAIL	Respond (go to link, reply, print coupon, etc.)	Propriety software used in-house	
FREE-STANDING INSERT (FSI) COUPONS	Redemption rate	In-house tabulation or coupon clearance services	

Measuring Success: Closing the Feedback Loop

For example, Roper Starch Worldwide measures whether a consumer "noted" an ad in a magazine. Burke, Inc. telephones households to track day-after recall for TV ads, providing reliable estimates of customer awareness of specific brands. Large firms realize that Burke's costly services are usually worth the expense. A large packaged goods firm, for example, that spends $10 million in television advertising for a new brand would consider it a wise investment to spend $100,000 on post-advertising awareness research to assess the effectiveness of the TV ad. To help evaluate the explosive growth of Internet advertising, the search giant Google provides common metrics that report the number of hits to a Web site. If you use a service by comScore, Inc., you can get a descriptive profile of those who searched for the product category or your competitors' Web sites.

Assessing the impact of advertising on buying behavior is often the communications goal of many marketing campaigns for convenience products such as those found in grocery or drug stores. In most of these product categories, consumers are already aware of the leading brands. In order to grow the brand, a firm must attract its competitor's customers. This is often accomplished by a price promotion. Coupon redemption programs have been a popular option for firms that distribute through retail channels. In addition, test marketing and simulated test marketing can provide realistic information on purchases of new product introductions.[206]

For B2B firms, the impact on sales is measured internally. This direct response to advertising can be tracked by customer calls to the firm's 800 number, or by tabulating reply cards or Web site logons. The firms will not only find out the number of prospects that the advertising attracted, but who they are.

In sum, there is no single approach to assess the impact of advertising delivered through the mainstream media, where there is no interaction or feedback metric built into the medium. This is why direct marketing methods—whether they are delivered in print as direct mail, or electronically through the Internet—are becoming more attractive to advertisers in this age of accountability. In the next section, we will provide a detailed example of the use of one metric, lifetime value.

DIRECT AND INTERACTIVE ADVERTISING: ROI AND LTV

Although direct marketing is usually a major facet of the marketing communications program of many firms, it has for a number of years been set apart from mass media communications planning. But with interactive communications, the impact of the advertisement can be determined immediately. Did a person open an email message? Did an Internet user click on a Google ad link? These metrics are common features on the dashboards of many campaign management software systems, as shown in Chapter 5.

Recent research reported by the CMO Council tabulated the top metrics marketers used to measure the success of personalized communications delivered by interactive or direct media. The top five metrics (multiple responses accepted) were:

- conversion and close rates (55%),
- email opening and forwarding rates (45%),
- Web site traffic and page views (42%),
- increased customer retention, reduced churn (41%), and
- quality and volume of leads (40%).

Customer value and profitability were used as metrics by 29% of respondents, and transaction frequency and size were used by 27%. The metrics least used were word-of-mouth referral and buzz (16%), and offer awareness and recognition levels (11%).[207]

While opened emails or Internet link click-through rates assess attention or interest in a brand, product, or retailer, it is the impact of the message on sales, customer value, and profitability that has been the metric of choice for direct marketers for decades. Below, we examine the metrics used to determine the return on interactive marketing communications investments across all media.

Media planning starts with cost per thousand (CPM), which is the cost to reach 1,000 people measured by audience size (broadcast), circulation (magazines and newspapers), or households that received direct mail. While this is a good starting point to assess the efficiency of different media choices, direct marketing specialists have developed an extensive array of metrics to gauge not only the revenue but also the profitability of a campaign.[208] Response per thousand (RPM) is a simple computation that captures the response, often measured in terms of inquiries, visits, or orders, related to the total number of people contacted.

$$\text{Response per thousand (RPM)} = \frac{\text{Total response (e.g., number of orders)}}{\text{Quantity contacted}/1000}$$

The cost per response (CPR) is determined by dividing the cost of contacting the audience by the total number that responded.

Cost per response (CPR) = CPM/RPM

For example, how will the owner of a new Thai restaurant in a suburban neighborhood of a small city choose between advertising its grand opening on cable TV or by direct mail? The media plan presented by an advertising agency would likely compare the cost per thousand (CPM) of the two media choices for reaching the target market of local, ethnic food lovers who like to dine out (see Table 8.3). In the first option, let's say a cable TV campaign requires $25,000 for producing the ad and $75,000 for broadcasting it to the 500,000 cable TV household subscribers in the region. The ad would be repeated 30 times during the week prior to the restaurant's opening event, and it would be shown on the Food Network, the Weather Channel, and during local news programs. Because the total cost is $100,000 for contacting 500,000 households, the CPM for this option would be $200. That is, for every 1,000 potential customers contacted, the restaurant owner would pay $200.

Compare this to a direct mail campaign. The advertising agency could buy a mailing list of the 20,000 homes within five miles of the new restaurant. A single mailing would appear in residents' mailboxes the week before the restaurant opens. If printing and mailing costs are $20,000, the CPM will be $1,000, or, for every 1,000 potential customers that the mailing will reach, the restaurant owner pays $1,000.

Table 8.3 *Comparison of two media choices for a hypothetical advertising campaign*

	PRODUCTION COSTS	MEDIA COSTS	NUMBER REACHED	CALCULATION OF CPM	CPM
CABLE TV	$25,000	$75,000	500,000	$100,000 / 500,000/1000	$200
DIRECT MAIL	$10,000	$10,000	20,000	$20,000 / 20,000/1000	$1,000

Which medium would you choose? If CPM is the most important consideration, you would obviously choose the cable TV ad because it delivers a lower CPM. Most mass media advertising will net a lower CPM than direct mailing methods. But a sophisticated media buyer is also concerned about the effectiveness of the advertising medium. A one-week, 30-slot exposure to 500,000 cable TV households will include a lot of wasted coverage. People who live 50 miles away from the restaurant may get the message, but won't be likely to drive that far to the grand opening. Perhaps, if some work nearby, they may consider going for lunch. But how many people would even notice the cable TV ad? With a direct mail campaign, on the other hand, patrons can be targeted by geographic location. However, would a single mailing be enough to get the word out or to initiate action?

In order to effectively compare media choices, a marketer needs to measure the outcomes of communication efforts. Taking our example one step further, let's imagine what happens at the weekend grand opening event. Table 8.4 gives the response per thousand (RPM) and cost per response (CPR) for an outcome of 100 dinner patrons as the result of each media campaign.

Table 8.4 *RPM and CPR for an outcome of 100 dinner patrons*

	CPM (COST PER M)	NUMBER OF PATRONS (SALES RESPONSE)	RPM (RESPONSE PER M)	CPR (COST PER RESPONSE) (CPM/RPM)
CABLE TV	$200	100	$\dfrac{100}{500{,}000/1000} = 0.2$	$\dfrac{\$200}{0.2} = \$1{,}000$
DIRECT MAIL	$1,000	100	$\dfrac{100}{20{,}000/1000} = 5$	$\dfrac{\$1{,}000}{5} = \$200$

Using the RPM metric, the direct mail campaign appears to be more efficient, since the owner would generate five orders per thousand people contacted using direct mail, versus generating only 0.2 orders per thousand people contacted with cable TV. But the CPR metric gives the best reason to favor direct mail. The restaurant will only have to spend $200 per order received with direct mail, versus spending $1,000 to get one order from the cable TV campaign.

While these metrics are handy measures that can deliver the accountability that business decisions require, they often only measure the effectiveness of a single campaign on a short-term outcome, as in the example above. To track buying behavior over time, the measure of *lifetime value* (LTV) has been adopted. LTV is defined as "the net present value of the profit that you will realize on the average new customer during a given number of years."[209] Championed by loyalty marketing consultants, LTV computations provide a metric to assess whether an initial investment in acquiring new customers will pay off in the long run. Marketers can also use LTV as a framework for comparing the value of the personalized printed communications used in direct mail advertising to the more common static or generic printed mailers.

For example, a fictional apparel catalog retailer called Great Northern wants to develop a new line of products that includes sports items such as golf, tennis, and skiing equipment, and target the new products to existing apparel customers. One approach to advertising this new line would be to develop a single catalog containing all of Great Northern's new sports equipment products, and then mail it to the company's list of apparel customers. The catalog would have static content and only the name and address of the customer would be printed on the cover for U.S. mail distribution. Let's say Great Northern sends such a common catalog to one million of its customers. In order to use an LTV computation for customers who respond during the first year, the revenues associated with each customer and the campaign costs must be calculated. Let's say Great Northern gives us the following data:

Table 8.5 *Hypothetical results from static/mail merge catalog campaign*

Response rate to new generic catalog (percent of catalog recipients who placed an order)	2%
Average number of new orders per year per customer	1.8
Size of order (average)	$450.00
Cost of goods sold, as a percentage of list price	70%
Cost of designing, printing and mailing the catalogs, per piece	$0.80
Discount rate (net present value, or NPV—the discounted value of money)	0.12

Using traditional direct marketing metrics, if the mailing goes to one million households and 20,000 customers place an order, the CPR (cost per response) is $40.00 ($800,000 to mail the catalogs / 20,000 responses) and the RPM (number of orders per thousand catalogs mailed) is 36 (2% x 1,000 x 1.8). Table 8.6 calculates the LTV of these 20,000 customers who ordered from the new catalog, using the above assumptions and applying NPV (net present value) in the first year. (Some practitioners argue that NPV should first be applied in the second year.)

This single promotion yields an NPV profit of $3.6 million in the first year. The LTV, though not a particularly useful metric in the first year, is $181.25 per customer acquired. The LTV metric becomes useful when we extend the timeframe to answer the question: What is the value of this customer base over time? If we assume that 60% of these customers who bought equipment in the first year buy again but gradually reduce the number of purchases they make over the next few years due to saturation or lifestyle changes, then we can calculate a four-year LTV (see Table 8.7).

The LTV per customer recruited in the first year has grown, by the fourth year, to $337. This information alone is not particularly useful but sets a bench-

Table 8.6 *LTV calculation for hypothetical static/mail merge catalog campaign*

METRICS	YEAR 1
CUSTOMER DATA	
# OF MAILINGS	1,000,000
RESPONSE RATE	2.0%
# OF CUSTOMERS ACQUIRED	20,000
# ORDERS/YEAR (AVERAGE)	1.8
REVENUE DATA	
AVERAGE ORDER SIZE	$ 450
TOTAL REVENUE PER YEAR	$ 16,200,000
COST OF GOODS SOLD (70% OF LIST PRICE)	$ 11,340,000
NET REVENUE	$ 4,860,000
MARKETING COSTS	
COST PER CATALOG (DESIGN, PRINTING AND POSTAGE)	$ 0.80
TOTAL MARKETING COST	$ 800,000
PROFIT	
NET PROFIT	$ 4,060,000
DISCOUNT RATE	1.12
NPV PROFIT	$ 3,625,000
AVG. CUSTOMER LTV (NPV PROFIT DIVIDED BY THE NUMBER OF CUSTOMERS ACQUIRED IN FIRST YEAR)	$ 181.25

mark against the measure of "what-if" scenarios for different campaigns.

Now let's compare this static mailing with a personalized catalog. Great Northern can customize its catalog based on the apparel purchasing behavior of its customers. For example, customers who purchased ski parkas can receive a ski equipment catalog, and those who purchased golf polos can receive a golf equipment catalog. We'll increase the hypothetical response rate in Table 8.8 to 5% because of the usual increased response rate to personalized campaigns, and we'll increase the cost of one catalog mailed to $2.00 because data-driven printing is more expensive than static printing.

Measuring Success: Closing the Feedback Loop

Table 8.7 Four-year LTV calculation for hypothetical static/mail merge catalog campaign

METRICS	YEAR 1	YEAR 2	YEAR 3	YEAR 4
CUSTOMER DATA				
RETENTION RATE	—	60%	60%	60%
# OF CUSTOMERS	20,000	12,000	7,200	4,320
# ORDERS/YEAR	1.8	1.5	1.2	1.0
REVENUE DATA				
AVERAGE ORDER	$ 450	$ 450	$ 450	$ 450
TOTAL REVENUE	$ 16,200,000	$ 8,100,000	$ 3,888,000	$ 1,944,000
COST OF GOODS SOLD (70%)	$ 11,340,000	$ 5,670,000	$ 2,721,600	$ 1,360,800
NET REVENUE	$ 4,860,000	$ 2,430,000	$ 1,166,400	$ 583,200
MARKETING COSTS				
COST PER MAILING	$ 0.80	$ 0.80	$ 0.80	$ 0.80
TOTAL MARKETING COST (BASED ON MAILING TO PRIOR YEAR'S CUSTOMERS)	$ 800,000	$ 16,000	$ 9,600	$ 5,760
PROFIT				
NET PROFIT	$ 4,060,000	$ 2,414,000	$ 1,156,800	$ 577,440
DISCOUNT RATE	1.12	1.25	1.40	1.57
NPV PROFIT	$ 3,625,000	$ 1,924,426	$ 823,387	$ 366,974
CUMULATIVE NPV PROFIT	$ 3,625,000	$ 5,549,426	$ 6,372,813	$ 6,739,787
AVG. CUSTOMER LTV (ON 20,000 CUSTOMERS ACQUIRED IN YEAR 1)	$ 181.25	$ 277.47 ($5,549,426/20,000)	$ 318.64	$ 336.99

Table 8.8 Results from hypothetical static and personalized catalog campaigns

METRICS	STATIC	PERSONALIZED
RESPONSE RATE	2%	5%
AVERAGE NUMBER OF NEW ORDERS PER YEAR PER CUSTOMER	1.8	1.8
SIZE OF ORDER	$450	$450
COST OF GOODS SOLD AS A PERCENTAGE OF LIST PRICE	70%	70%
COST OF DESIGNING, PRINTING AND MAILING EACH CATALOG	$0.80	$2.00
DISCOUNT RATE	0.12	0.12

Using the traditional direct marketing metrics, the CPR for the personalized catalog campaign remains at $40.00 ($2,000,000 to mail the catalogs / 50,000 responses), but the RPM has risen to 90 (5% x 1,000 x 1.8). In other words, while the CPR and LTV values remain the same during the first year, the number of orders has substantially increased. As shown in the right-hand column of Table 8.9, the new campaign will yield a NPV profit of just over $9 million. When compared to the results from the static catalog campaign, customization is expected to produce a larger total profit.

Most business managers would quickly choose the campaign that generated $9 million as opposed to a plan that generated $3.6 million. The simple direct marketing metric of RPM might also lead them to the same conclusion.

The LTV calculations will add insight to the effectiveness of each campaign over time because LTV assesses the future impact of an original customer group. For the static catalog, if we assume that 60% of the customer group continues buying but makes fewer purchases every year, then after four years, 4,320 of the original 20,000 would remain as "active buyers." The LTV of each of these customers would be $336 (refer back to Table 8.7).

However, the personalized catalog campaign will give us different numbers, since personalization efforts usually pay off with a higher retention rate over time and an increase in the average number of orders per year. As shown in Table 8.10, after four years, an estimated 16,800 of the original 50,000 customers would remain "active buyers" in the sports division of Great Northern. The LTV of each of these customers would be $465.

In theory (that is, if you accept all of these assumptions), the investment in personalized marketing communications pays off both in the short

Table 8.9 One-year LTV calculation of hypothetical static and personalized catalog campaigns

METRICS	STATIC/MAIL MERGE CATALOG	PERSONALIZED CATALOG
CUSTOMER DATA		
# OF MAILINGS	1,000,000	1,000,000
RESPONSE RATE	2.0%	5.0%
# OF CUSTOMERS WHO ORDERED	20,000	50,000
# ORDERS/YEAR (AVERAGE)	1.8	1.8
REVENUE DATA		
AVERAGE ORDER SIZE	$ 450	$ 450
TOTAL REVENUE PER YEAR	$ 16,200,000	$ 40,500,000
COST OF GOODS SOLD (70% OF LIST PRICE)	$ 11,340,000	$ 28,350,000
NET REVENUE	$ 4,860,000	$ 12,150,000
MARKETING COSTS		
COST PER MAILING (PRINTING AND POSTAGE)	$ 0.80	$ 2.00
TOTAL MARKETING COST	$ 800,000	$ 2,000,000
PROFIT		
NET PROFIT	$ 4,060,000	$ 10,150,000
DISCOUNT RATE	1.12	1.12
NPV PROFIT	$ 3,625,000	$ 9,062,500
AVG. CUSTOMER LTV (FOR CUSTOMERS WHO ORDERED IN YEAR 1)	$ 181.25 ($3,625,000/20,000)	$ 181.25 ($9,062,500/50,000)

run, as demonstrated by RPM, and in the long run as demonstrated by LTV. Figure 8.1 shows the improving gains year-by-year in the LTV of the personalized offer.

The marketing metrics we have discussed are extremely useful in assessing the impact of promotional tactics on sales and profits. And, as we have also seen, demonstrating the effects of personalized marketing communications is necessary for any business considering whether to implement such a program.

Table 8.10 Four-year LTV calculation for hypothetical personalized catalog campaign

METRICS	YEAR 1	YEAR 2	YEAR 3	YEAR 4
CUSTOMER DATA				
RETENTION RATE		60%	70%	80%
# OF CUSTOMERS	50,000	30,000	21,000	16,800
# ORDERS PER YEAR	1.8	2	2.2	2.4
REVENUE DATA				
AVERAGE ORDER	$450	$450	$450	$450
TOTAL REVENUE	$40,500,000	$27,000,000	$20,790,000	$18,144,000
COST OF GOODS SOLD (70% OF LIST)	$28,350,000	$18,900,000	$14,553,000	$12,700,800
NET REVENUE	$12,150,000	$8,100,000	$6,237,000	$5,443,200
MARKETING DATA				
COST PER MAILING	$2.00	$2.00	$2.00	$2.00
TOTAL MARKETING COST	$2,000,000	$100,000	$60,000	$42,000
PROFIT				
NET PROFIT	$10,150,000	$8,000,000	$6,177,000	$5,401,200
DISCOUNT RATE	1.12	1.25	1.40	1.57
NPV PROFIT	$9,062,500	$6,377,551	$4,396,667	$3,432,560
CUMULATIVE NPV PROFIT	$9,062,500	$15,440,051	$19,836,718	$23,269,278
AVG. CUSTOMER LTV	$181.25	$308.80	$396.73	$465.39

Measuring Success: Closing the Feedback Loop

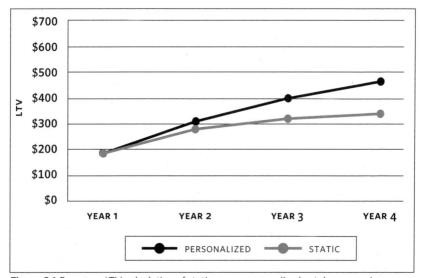

Figure 8.1 *Four-year LTV calculation of static versus personalized catalog campaigns*

CONCLUSION

To make a business case for personalized communications, the media consultant must be able to demonstrate that a personalized offer will deliver more profits over time than a larger quantity of mass-produced offers. To do this, the feedback loop for marketing communications planning must be articulated at the beginning of the communications planning process. Data must then be captured, and a system put in place to handle it. There are two ways to do this: outsource the whole process, or use internal resources. The nature of the firm and the products it markets will likely influence which alternative is more appropriate.

For convenience goods, where mass media advertising is the primary medium used, outsourcing the process makes sense. One drawback to this alternative is that the buying behavior of individual customers or prospects will not necessarily be evident. If a firm decides to handle its data in-house, it can also easily track individual buying behavior. While this often requires a large investment in IT infrastructure, such as an enterprise-wide CRM system or campaign management system, having these resources will pay off in the long run because the firm will then be able to take advantage of personalized marketing possibilities. Whether a firm decides to measure the impact of marketing efforts internally or by outsourcing methods, an orderly process for creating marketing campaign feedback systems needs to be followed.

The first step, of course, is to define the marketing objectives. Next, a means through which the prospect or customer responses will be captured must be created, including how these data will be reported. This could entail setting up and staffing an 800-number, asking for specific information on a Web site, or including a postage-paid reply card on a general mailing. Who will be assigned the responsibility to track results? As we saw above, this can be outsourced to an external vendor (for example, an ad agency or direct marketing firm), or be managed by an internal sales support staff. In either case, the database infrastructure that gathers the responses (call center, Web site, etc.) should be integrated with the marketing and fulfillment database systems, which will establish the information distribution protocols. The pertinent questions are what information needs to be reported, to whom, and how often.

In closing, in order for most resource-intensive, personalized marketing communications tactics to be championed by either an internal marketing manager or an external communications consultant, the efficiency of the tactic must be demonstrated. Whether the manager or consultant uses simple direct marketing measures (such as RPM), or more sophisticated LTV computations (where individual consumer behavior is tracked over time), the feedback system must be a part of the initial plan.

NOTES

196. Quotations Page, "Quote Details: John Wanamaker: Half of the money I..."
197. Strategic Planning Institute, "SPI Home."
198. Barwise and Farley, "Which Marketing Metrics Are Used and Where?"
199. Winer, "What Marketing Metrics Are Used by MSI Members?"
200. Wyner, "The ROI Toolkit."
201. Duffy, "Kraft's Return on Marketing Investment: Portfolio Management Planning Implications."
202. Brady, Kiley, and bureau reports, "Making Marketing Measure Up: The Pressure Is On to Take the Guesswork Out of Ad Spending;" Beirne and Hein, "Marketer's Mantra: It's ROI or I'm Fired!"
203. McGuire, "An Information Processing Approach to Advertising Effectiveness."
204. Weilbacher, "Point of View: Does Advertising Cause a 'Hierarchy of Effects?'"
205. Schultz, "Determine Outcomes First to Measure Efforts."

206. Clancy, Shulman, and Wolf, *Simulated Test Marketing*.
207. CMO Council, "The Power of Personalization: The Impact and Influence of Individualized Content Delivery."
208. Nash, *The Direct Marketing Handbook*.
209. Hughes, *Strategic Database Marketing: The Masterplan for Starting and Managing a Profitable Customer-Based Marketing Program*.

CHAPTER NINE

THE INTELLIGENT USE OF PERSONALIZATION

THE FOCUS OF THIS BOOK has been to help marketing decision makers and their advisers understand the strategy and technology requirements to create personalized promotional materials that are welcomed by the receiver. We have identified these steps to create effective personalized marketing campaigns:

1. Find a personalization tactic that fits the business strategy.
2. Use or acquire knowledge about customers in order to personalize messages.
3. Develop a creative campaign (which often includes market tests).
4. Implement the campaign.
5. Assess its effectiveness.

To get started, a firm should begin with simple levels of personalization. With subsequent organizational learning about customers through data capture and analysis, more creative personalized approaches can be designed as a result of insights gathered through data mining. Upon implementation, the success of the communications should be tracked using direct marketing methods. If the LTV metric is used and validated, it can also help the firm complete "what if" scenarios in planning for future marketing investments. These data will then provide a solid business case for incremental investments.

But for firms that are trying to determine whether they should initiate a personalized campaign, there is a bit of a chicken-and-egg challenge: How can you show the effectiveness of personalization in order to get the commitment of upper management to invest in it if you have never conducted such a campaign? Many marketing executives use experienced communications consultants such as advertising agencies or direct marketing firms to convince top management that these programs are likely to be successful. The critical issue to demonstrate is whether this tactic will enable a firm to achieve its business objectives. Will the investment in time and money pay off, and will the opportunity costs be worth the trouble?

The purpose of this final chapter is to address the likely questions that top management will ask of communications consultants who want to sell personalized marketing communication to a firm that has never used it before. First, can the communications consultant demonstrate unequivocally that the program will improve the bottom line? Second, will these programs grow long-term customer value? Let's examine the evidence in the public domain.

EVIDENCE FOR THE ASSUMPTIONS

First, has it been proven that response rates go up when a company uses personalized advertising? The short answer is "Yes." The most convincing research regarding the difference between personalized and static mailings comes from studies that use experimental methods (that is, that include both experimental and control groups). In a 2004 report from PIA/GATF's Digital Printing Council (DPC), the response rate for personalized color direct mail campaigns ranged from 6% to 75%, with an average of 21%.[210] The response rates were, on average, 5.6 times higher for personalized color versions than for simple mail-merge applications. For example, a German utility firm mailed a personalized brochure to one sample of customers and a generic brochure to a second sample. In both mailings, the brochure described the benefits of a new energy savings program, but the personalized brochure included the estimated energy savings for each household based on its own particular history of energy use. The average inquiry rate for the personalized brochure was 6.3%, versus only 2.4% for the generic brochure. The personalized mailing delivered a significantly higher response rate than the generic one.

More recent research also supports these earlier conclusions. Using a similar experimental design as those above, Staples Business Depot tested static versus personalized letters sent to lapsed customers, numbering over 35,000 accounts.[211] One half of the list received a standard letter without any personalized content and the other half received a similar-looking letter but the content was customized to each receiver. Personalized elements included written details of a gift for the reciepent, which varied by past purchase history, a description of the current loyalty program the recipient was registered for, and a coupon that was related to past purchases. The letter also included barcoding to track responses. The results revealed that the group that received the personalized letter yielded a 50% lift in response and a 37% rise in profit per customer.[212] This "plain vanilla" letter showed the impact of personalization that can be delivered without flashy graphics or color printing.

Other research has used survey methods to capture the impact of personalization. In the 2008 CMO research study cited previously, 56% of respondents indicated that their personal communications investments performed better than traditional mass market delivery.[213] In terms of broadly-defined "effectiveness," the executives reported that the top three most effective personalized communications activities were:[214]

- individualized emails and letters (55% reported as effective),
- targeted database marketing, leveraging personal profiles (35%), and
- opt-in, permission-based marketing programs (30%).

These three tactics were also the ones most frequently used by the respondents. Other tactics received lower levels of reported effectiveness, but these results should be taken in the context of the relatively smaller base that used them. Table 9.1 presents tactic effectiveness in relation to frequency of use in the CMO study.

Table 9.1 *Effective personalized communications tactics*[215]

TACTIC	PERCENTAGE WHO USE	% REPORTING TOP 3 IN EFFECTIVENESS
PRINT ON-DEMAND COLLATERAL INCORPORATING PERSONALIZED CONTENT	31%	18%
PERSONALIZED URLS (INTERNET DOMAINS)	26%	16%
USE OF COOKIES AND SEARCH TRACKING TECHNOLOGY	25%	13%
CONSUMER-CUSTOMIZED WEB SITES (USER CONTROLS)	15%	6%
WEB SITE PAGE DELIVERY BASED ON SEARCH HISTORY	11%	6%
TRANSPROMO COMMUNICATIONS	9%	4%

The table reveals that about half of those who used any one tactic reported it as one of the top three in effectiveness.

Evidence of personalization building customer loyalty is harder to find. Numerous anecdotes and case studies have been reported by business consultants who tout the pay-off for investing in loyalty programs.[216] But the real question for communications consultants who are trying to sell personalization is this: Can marketers build loyalty by using technology-mediated communications instead of personal interactions? As discussed in Appendix 4A, loyalty and relationship marketing programs are usually built on the foundations of B2B exchanges that rely on frequent face-to-face communications. In the retail sector, interpersonal communications are also at the heart of what many consumers include in their definitions of "relationship."[217] Can we build relationships and loyalty with personalized communications instead of these face-to-face communications? Or will the "junk mail" and "spam" characterizations of unwanted communications continue to haunt this most recent embodiment?

One 2001 study gives direct marketers some hope that personalized marketing communications can be used to build relationships. De Wulf, Odekerken-Schröder, and Iacobucci investigated whether relationship marketing tactics used by retailers would impact the degree to which consumers would perceive that a firm was "investing" in the relationship.[218] They wondered whether the perception of an improved relationship investment (as evidenced by higher relationship quality perceptions) would lead to increased feelings of loyalty and repeat purchases. Using food and clothing retailers in both Europe and the U.S., the researchers assessed the impact of four marketing tactics:

- direct mail,
- interpersonal communication,
- preferential treatment, and
- tangible rewards.

Interpersonal communication had the largest impact of the four tactics on building relationship quality perceptions. This was the case for both societies. Direct mail also contributed to the perceptions of higher relationship quality, *but only for the European customers*. That is, direct mail did not improve perceptions of relationship quality for U.S. customers.[219]

However, there were differences among U.S. customers in their perceptions of relationship quality and in their repeat buying patterns according to their degrees of product involvement and "relationship proneness." That is, those who were more involved with a product and who were more likely to acknowledge that they had a relationship with a commercial entity did respond more often to direct mail. While this may suggest that personalized marketing communications can contribute to building loyalty and LTV, these are the conclusions of just one study. Clearly, there is a need for more research in this area.

If the results of this study are validated with similar findings, it would suggest that in order for firms to build customer relationships using personalized communications, they must first understand the attitudes that customers have towards them. As we saw in Chapter 4, firms must create a dialogue with customers to understand their preferences regarding how they want to be treated by commercial entities.

OTHER APPLICATIONS FOR PERSONALIZATION

While the marketing and advertising world awaits the outcome of additional research, marketers can think broadly about the value of a variety of personalized communications options. Building connections with customers extends beyond the social exchange notions of forming a relationship. Market researchers Rust, Zeithaml, and Lemon view *customer equity* as the basis for a new strategic framework from which to build customer-centered programs that are financially accountable.[220] They define a firm's customer equity as the sum of the discounted LTV of all its customers, and count three ways to build it:

- value equity,
- brand equity, and
- retention equity.[221]

Personalization can be an effective tactic in all three routes to building customer equity. Table 9.2 shows how personalization can be applied in each case.

Table 9.2 *Three ways to build customer equity*[222]

EQUITY TYPE	DEFINITION	PERSONALIZATION GOAL
VALUE	A customer's objective assessment of the utility of a brand, driven by the product's quality, price, and convenience.	Right time, right place, right offer that leads to preference
BRAND	A customer's subjective and intangible assessment of a brand built through image and meaning. This assessment is influenced by brand awareness and "corporate citizenship."	Build a depth of knowledge in customers about the firm's brands that leads to commitment
RETENTION	Tendency of the customer to "stick with" a brand above and beyond the objective and subjective assessments.	Repeat buying behavior

In order for *value* equity to accrue, the customer must regard the firm's offering as the best value among all competitive offers. A customer who holds this view will continue to buy from the firm as long as its brand continues to deliver the best value. Simple personalization programs can reinforce this customer assessment. Tactics such as reminder postcards serve this function well. A customer who is reminded by a car dealer's auto service center to take care of some routine car maintenance through an attention-getting mailing might otherwise have neglected to call for service. If the message is timely and helps the consumer keep track of annoying responsibilities, then it delivers value and contributes to the perceived superiority of one brand over another.

To build *brand* equity, the business goal is to create an emotional connection between the customer and the brand. This approach works for high-involvement and lifestyle products. For example, some automobile manufacturers create customized publications for their customers, with one such manufacturer being Audi. Audi owners get the Audi magazine once a year. The slick, expensive-looking magazine arrives at their homes because Audi believes that most luxury or performance car owners have an emotional connection to their cars. Audi tries to bolster this connection with articles about Audi engineers, car races that Audis have won, and the exotic and beautiful vacation spots they can tour in their Audis. High-involvement products (those that entail a significant financial or social risk) attract the type of consumers who will develop emotional attachments that can be fueled with more information. These consumers will repurchase the same or similar products because of their affective connection to the brand or manufacturer. In these best-case scenarios, a consumer's commitment to brand is so complete that alternative options are not considered.

Many brand-building efforts also have an elaborate Internet community presence that promotes interactivity. For example, the Jeep community hangs out at http://www.jeep.com/en/experience/community/urban_ranger/. Here Jeep owners can find pictures and videos of the latest events that they might have attended—and show their friends! There is even a Jeep YouTube channel.

All of this interactivity comes at a price. Maintaining a Web presence may be expensive (never mind planning and staffing Jeep Jamboree events), but the service expectations of these loyal customers are also very high. These brand "fans" expect immediate responses to inquiries, whether they are questions about new features, reports of problems with the product, or simply account or balance information requests. Firms that pursue the route to customer equity must invest in an information infrastructure that is enterprise-wide, connecting every customer *touch point* within the firm. The only limitation to leveraging the customer data accumulated by CRM activities is the creativity of the marketing or brand manager.

The third route to customer equity is *retention* equity. The continued patronage of customers may result from inertia (as in the repeat purchasing of convenience products) or the desire for rewards or savings (as in frequent flyer programs). Saving on everyday items is valued by grocery store shoppers, and is delivered when a "frequent buyer card" is scanned at the point of purchase. Though an enormous amount of data is gathered by retailers using these programs, few of them apply it to direct, personalized communications with customers because of the privacy concerns of shoppers. The next case study describes the effective use of custom publishing by one large food retailer that targets its frequent shoppers.

When the cost of switching to another product or service provider is high, most customers will, again because of inertia, repurchase from the same provider. Rewards and preferential treatment provide incentives to maintain a customer's continued patronage. These are often delivered by mail, perhaps on or with a monthly statement, to encourage customers to cash in on their "preferred status" through additional patronage. The preferential treatment program for Harrah's casino patrons discussed earlier in this book not only made Harrah's the logical choice for repeat business, but also bestowed special privileges and recognition on patrons when they did return. However, these retention tactics are easily copied by competitors. It is not unusual for business travelers, for example, to have multiple frequent flier cards.

All of these examples suggest that many firms can use personalized communications to achieve a variety of marketing goals. Regardless of the marketing strategy, personalization can be a vital way to build connections with customers. However, as personalization becomes more common, the attention-getting pull of simple forms of personalization may diminish. For the future growth of this medium, insights from data mining must be combined with creative marketing messages to ensure success. Thorough customer knowledge will be required to render personalization an effective marketing tool. The steps and strategies presented in this book will help print and interactive media services providers consult with their clients to capture and creatively use customer information to fulfill the promise of personalization.

CASE STUDY: CUSTOM PUBLISHING

Custom publishing is not a new phenomenon. John Deere first published the magazine *Furrow* for farmers and ranchers in 1895.[223] It is still in circulation today, connecting the equipment manufacturer with its dealers and helping them deliver more value to their own local customers. The magazine's circulation peaked at 5 million in the early part of the 20th century, and now has a worldwide circulation of more than 1.5 million. It is customized by geographic area, with nine regional editions in the U.S. and three in Canada.

Corporations are waking up this medium. According to the trade group American Business Media, there were approximately 50,000 customized publications produced annually in the U.S in 2003.[224] Newsletters were the most popular category at 53%, and 24% were customized magazines. E-publications were growing at a fast rate and then comprised 17% of the overall annual production. Corporations spend over $20 billion annually to print and distribute these vehicles[225]. Not only do they provide a means for firms to connect to their customers, these publications also advertise products. The revenue stream these publications generate can offset their own printing and distribution costs.

A great example is *Menu*, a magazine published by Wegmans Food Markets in Rochester, NY.[226] *Menu* was started in 2001 and comes out four times per year. The magazine's goal is clearly stated on its cover: "Helping you make great meals easy." In addition to menus and food preparation techniques, the magazine features stories about Wegmans' merchandising team members, department managers, and executive chefs, many of whom are Culinary Institute of America graduates. Wegmans teams travel the world to find the ingredients and meal ideas that their shoppers want. Because the focus is on these in-depth feature stories, the number of advertisements is low.

Of the hundreds of thousands of copies that are printed, over half are mailed to a selection of "Wegmans Shopper's Club" loyalty card members. The remaining copies are placed in stores and are available to shoppers at a $4.00 cover price. The magazine is designed internally and printing is outsourced to a major commercial printer. The contents of the magazine, including all recipes, are also featured on the company's website. Former customers who have moved out of the Northeast (where Wegmans Food Markets are located) can also order current copies or back issues of *Menu* on the site.

Judging by customer feedback gathered by the company's consumer affairs office, periodic satisfaction surveys, and the increase in the sales of featured items in the magazine, *Menu* has successfully connected grocery shoppers to the Wegmans brand. The content is not customized for individuals (except for occasional small-scale tests), but the magazine is nevertheless a good example of a printed publication targeted to a firm's high-value customers for the purpose of strengthening its bond with them.

NOTES

210. INTERQUEST, *Opportunities in Color Variable Data Imaging.*
211. Tolliver-Nigro, "Side-by-Side Test Proves 1:1."
212. Ibid.
213. CMO Council, "The Power of Personalization: The Impact and Influence of Individualized Content Delivery."
214. Ibid.
215. Ibid.
216. Reichheld, *The Loyalty Effect: The Hidden Force behind Growth, Profits, and Lasting Value.*
217. Sorce and Edwards, "Defining Business-to-Consumer Relationships: The Consumer's Perspective."
218. De Wulf, Odekerken-Schröder, and Iacobucci, "Investments in Consumer Relationships: A Cross-Country and Cross-Industry Exploration."
219. See Dash, "Europe Zips Lips; U.S. Sells ZIPs," for a discussion of how privacy and information concerns differ between the U.S. and Europe.
220. Clancy and Krieg, "Customer Equity: A Fix for Modern Marketing."
221. Rust, Zeithaml and Lemon, *Driving Customer Equity: How Customer Lifetime Value is Reshaping Corporate Strategy.*
222. Ibid.
223. John Deere Web site, *The Furrow.* http://johndeere.com/en_US/ag/furrow/index.html.
224. Publishing & Media Group, *Custom Publishing: Opportunities Abound for B-2-B Publishers.*
225. Ibid
226. John Hawkes, (editor of *Menu* magazine, a publication of Wegmans Food Markets, Inc.)

REFERENCES

"45th Annual Print Design Survey." *Graphic Design USA*, June 2008. http://gdusa.com/issue_2008/06_jun/feature/June_08_Print_Survey.pdf (accessed September 15, 2008).

"2007 PIA/GATF Ratios Show Printing Industry Profits Increasing, PIA/GATF Press Release." *Printing News*, September 26, 2007. http://www.printingnews.com/web/online/Industry-News/2007-PIAGATF-Ratios-Show-Printing-Industry-Profits-Increasing/1$2923 (accessed June 17, 2008).

American Press Institute. "Newspaper Next: Blueprint for Transformation." http://www.newspapernext.org/2005/09/report_availability_1.htm (accessed March 14, 2009).

American Society of Magazine Editors. "Number of Magazine Titles, 1988–2007." *Editorial Trends & Magazine Handbook*. http://www.magazine.org/asme/editorial_trends/1093.aspx (accessed March 14, 2009).

Audit Bureau of Ciculations. "Print Circulation Trends," in Audience-FAX eTrends Tool. http://abcas3.accessabc.com/audience-fax/default.aspx (accessed September 15, 2008).

Bagozzi, R.P. "Toward a Formal Theory of Marketing Exchange." In *Conceptual and Theoretical Developments in Marketing*, edited by O.C. Ferrell, Stephen W. Brown, and Charles W. Lamb, Jr. Chicago, IL: American Marketing Association, 1979: 431–47.

Bartz, Diane. "Consumer Groups Urge 'Do Not Track' Registry." Reuters. April 15, 2008. http://www.reuters.com/article/governmentFilingsNews/idUSN1520070020080415 (accessed April 16, 2008).

Barwise, Patrick, and John U. Farley. "Which Marketing Metrics Are Used and Where?" *2003 MSI Reports* (Marketing Science Institute). Report no. 03-111, 2003. http://www.msi.org/publications/publication.cfm?pub=649 (accessed March 9, 2009).

Beirne, Mike, and Kenneth Hein. "Marketers' Mantra: It's ROI, or I'm Fired!" *Brandweek* 45, no. 37 (December 13, 2004): 14-15.

Brady, Diane, David Kiley, and bureau reports. "Making Marketing Measure Up: The Pressure Is On to Take the Guesswork Out of Ad Spending." *Business Week* 3912 (December 13, 2004): 112.

Briggs, Rex, and Greg Stuart. *What Sticks: Why Most Advertising Fails and How to Guarantee Yours Succeeds*. Chicago, IL: Kaplan Publishing, 2006.

Bronner, Fred, and Peter Neijens. "Audience Experiences of Media Context and Embedded Advertising: A Comparison of Eight Media." *International Journal of Market Research* 48, no. 1 (2006): 81-100.

Budington, Jon. Personal communication, September, 2008.

Calder, Bobby J., and Edward C. Malthouse. "Qualitative Effects of Media on Advertising Effectiveness." Northwestern (University) Media Management Center. http://www.medill.northwestern.edu/WorkArea/showcontent.aspx?id=70551 (accessed September 15, 2008).

Campanelli, Melissa. "Ace Hardware Insert Test Passes with Flying Colors." *DM News: The Online Newspaper of Record for Direct Marketers* 26, no. 43 (November 15, 2004).

———. "Credit Card Solicitations at Highest in '04." *DM News: The Online Newspaper of Record for Direct Marketers* 27, no. 12 (March 28, 2005): 6.

Cappo, Joe. *The Future of Advertising*. Chicago: McGraw Hill, 2003.

Clancy, Kevin J. Robert S. Shulman and Marianne M. Wolf. *Simulated Test Marketing: Technology for Launching Successful New Products*. New York: Lexington Books, 1994.

Clancy, Kevin J, and Peter Krieg. "Customer Equity: A Fix for Modern Marketing." *News Room*, Copernicus Marketing Consulting, February 20, 2001. http://www.copernicusmarketing.com/about/docs/customer_equity.htm (accessed on October 8, 2005).

CMO Council. "The Power of Personalization: The Impact and Influence of Individualized Content Delivery." http://www.cmocouncil.org/resources/form_pop.asp (accessed March 9, 2009).

Coviello, Nicole E., Roderick J. Brodie, Peter J. Danaher, and Wesley J. Johnston. "How Firms Relate to Their Markets: An Empirical Examination of Contemporary Marketing Practices." *Journal of Marketing* 66, no. 3 (July 2002): 33–46.

Creamer, Matthew. "Think Different: The Web's Not a Place to Stick Your Ads." *Advertising Age* 79, no. 11 (March 17, 2008): 3.

Cummings, Twyla, Howard Vogl, Claudia Cristina Alvarez Casanova, and Javier Rodriguez Borlado. *An Examination of Business and Workflow Models for U.S. Newspapers* (PICRM 2009-02). Rochester, NY: Rochester Institute of Technology, Printing Industry Center, January 2009.

Dahlén, Micael. "The Medium as a Contextual Cue: Effects of Creative Media Choice." *Journal of Advertising* 34, no. 3 (2005): 89–98.

Dash, Eric. "Europe Zips Lips; U.S. Sells Zips." *New York Times*, August 7, 2005, Section 4:1.

Davis, Ronnie H. "Update on Printing Industry Restructuring: Long-and-Short-Term Trends and Their Competitive Implications." *Economic and Print Market Flash Report* (Pittsburgh, PIA/GATF), April 17, 2008.

———, and Ed Gleeson. *Expanding the Print Market Space: Printers' Diversification into Ancillary Services.* Pittsburgh: PIA/GATF Press, 2008.

De Wulf, Kristof, Gaby Odekerken-Schröder, and Dawn Iacobucci. "Investments in Consumer Relationships: A Cross-Country and Cross-Industry Exploration." *Journal of Marketing* 65, no. 4 (October 2001): 33–50.

Dewitz, Adam. *Web-Enabled Print Architectures* (PICRM-2008-06). Rochester, NY: Rochester Institute of Technology, Printing Industry Center, 2008.

Direct Marketing Association. *2007 Response Rate Trends Report*. New York: Direct Marketing Association, 2007.

———. *Statistical Fact Book*. New York: Direct Marketing Association, 2007.

———. *The State of Postal and E-mail Marketing 2002: New List Trends and Results*. New York: Direct Marketing Association, 2002.

Drozdenko, Ronald G., and Perry D. Drake. *Optimal Database Marketing*. Thousand Oaks, CA: Sage Publications, Inc., 2002.

Duffy, Michael D. "Kraft's Return on Marketing Investment: Portfolio Management Planning Implications." In *Marketing Metrics*, Report No. 00-119, edited by Marion Debruyne and Katrina Hubbard. Summarized proceedings of Marketing Science Institute conference. Toronto, Canada, October 5-6, 2000: 33-36.

eMarketer, Inc. "Is Behavioral Targeting Bothersome?" April 11, 2008. http://www.emarketer.com/Articles/Print.aspx?id=1006175&src=print_article_graybar_article&xsrc=print1_articlex (accessed April 15, 2008).

Federal Communications Commission. "FCC Adopts 13th Annual Report to Congress On Video Competition and Notice of Inquiry for the 14th Annual Report." FCC news release, November 27, 2007. http://hraunfoss.fcc.gov/edocs_public/attachmatch/DOC-278454A1.pdf (accessed July 10, 2008).

Fenton, Howie. "Profit Generator?" *NAPL Business Review* 3, no. 2 (2008): 48-51.

Forrester Research, Inc. "The Digital Transformation." For American Business Media, October 2007. http://www.americanbusinessmedia.com/abm/ABMIntellResearch2007.asp?SnID=626885515 (accessed July 11, 2008).

———. "The B2B Digital Marketing Shift." For American Business Media, 2005. http://www.americanbusinessmedia.com/images/abm/ppt/ABM_Sales_Presentation_Final_v2_ext.ppt (accessed September 15, 2008).

Fournier, Susan, Susan Dobscha, and David Glen Mick. "Preventing the Premature Death of Relationship Marketing." *Harvard Business Review* 76, no. 1 (January/February 1998): 42–51.

Godin, Seth. *Permission Marketing: Turning Strangers into Friends and Friends into Customers.* New York: Simon and Schuster, 1999.

Gilmore, James H., and B. Joseph Pine, II. "The Four Faces of Mass Customization." *Harvard Business Review* 75 (January/February 1997): 91–101.

Green, Jeff. "In the Mail (or Dumpster)." *Cards & Payments* 19, no.5 (May 2006): 6.

Harrington, Richard J., and Anthony K. Tjan. "Transforming Strategy One Customer at a Time." *Harvard Business Review* 86, no. 3 (2008): 62–72.

Hotchkiss, Gord. "Personalization Catches The User's Eye." MediaPost Publications, September 13, 2007. https://www.mediapost.com/publications/index.cfm?fuseaction=Articles.showArticle&art_aid=67356 (accessed March 9, 2009).

Hughes, Arthur M. *Strategic Database Marketing: The Masterplan for Starting and Managing a Profitable Customer-Based Marketing Program.* 2nd ed. New York: McGraw-Hill, 2000.

Iacobucci, Dawn, and Johnathan D. Hibbard. "Toward an Encompassing Theory of Business Marketing Relationships (BMRs) and Interpersonal Commercial Relationships (ICRs): An Empirical Generalization." *Journal of Interactive Marketing* 13, no. 3 (1999): 13–33.

Industry Measure. "Variable Data Printing/1:1 Personalization: 2007." The Association of Graphic Solutions Providers (IPA). http://www.ipa.org/knowledge-center/industry-measure/variable-data-printing-11-print-personalization-2007 (accessed March 9, 2009).

InfoTrends, Inc. *The Future of Mail 2006: Direct Mail, Transaction, and "Transpromotional" Documents.* Weymouth, MA: InfoTrends, 2006.

——. *Multi-Channel Communications: The Content Publishing Workflow Challenge.* Weymouth, MA: InfoTrends, 2006.

——. *Trans Meets Promo... Is It More than Market Hype?* Weymouth, MA: InfoTrends, September 2008.

INTERQUEST, for the Digital Printing Council (DPC) of PIA/GATF. *Opportunities in Color Variable Data Imaging.* Alexandria, VA: Printing Industries of America (PIA) / Graphic Arts Technical Foundation (GATF), 2004.

Japsen, Bruce. "Patient Data Faced Exposure: Wellpoint Client Records Left Open to Possibe Theft." *Chicago Tribune*, April 16, 2008, business section: 1. http://pqasb.pqarchiver.com/chicagotribune/access/1462659741.html?dids=1462659741:1462659741&FMT=ABS&FM (accessed March 10, 2009).

Jayachandran, Satish, Subhash Sharma, Peter Kaufman, and Pushkala Raman. "The Role of Relational Information Processes and Technology Use in Customer Relationship Management." *Journal of Marketing* 69, no. 4 (October 2005): 177–92.

Jeep Web site. "Jeep. Community, Home Base." *http://www.jeep.com/en/experience/community/urban_ranger/* (accessed March 10, 2009).

John Deere Web site. *The Furrow.* http://www.deere.com/en_US/ag/furrow/index.html (accessed March 10, 2009).

Kalyanam, Kirthi, and Monte Zweben. "The Perfect Message at the Perfect Moment." *Harvard Business Review* 83, no. 11 (November 2005): 112–20.

Kane, Michael G. "Sprinting a Marathon: The Newspaper Transformation Today." Paul & Louise Miller Lecture, Rochester, NY: Rochester Institute of Technology, April 30, 2008. http://print.rit.edu/events/miller/kane/ (accessed March 15, 2009).

Kelley, Larry D., and Donald W. Jugenheimer. *Advertising Media Planning: A Brand Management Approach.* Armonk, NY: M.E. Sharpe, Inc., 2004.

Kotler, Philip. *Marketing Management.* Upper Saddle River, NJ: Prentice Hall, 2003.

"The Largest In-Plants." *In-Plant Graphics* 57, no.12 (December 1, 2007): 28, 32. http://www.ipgonline.com/article/weve-listed-largest-in-plants-according-annual-sales-number-full-time-employees-83668_1.html (accessed March 13, 2009).

Loveman, Gary. "Diamonds in the Datamine." *Harvard Business Review* 81, no. 5 (May 2003): 109–13.

Luo, Xueming, and Naveen Donthu. "Benchmarking Advertising Efficiency." *Journal of Advertising Research* 41, no. 6 (November/December 2001): 7–18.

Macro, Kenneth. "In-plants: The Next Generation." *In-Plant Graphics* 57, no. 3 (March 2007): 28–32.

Magill, Ken. "Shop.org Projects Online Sales Will Rise 17% in 2008." *Direct*, April 8, 2008. http://directmag.com/news/online-sales040808/index.html (accessed April 17, 2008).

Mahajan, Vijay, Eitan Muller, and Yoram Wind. (Eds.) *New-Product Diffusion Models.* Boston: Kluwer Academic Publishers, 2000.

Marketing Charts. "Catalog Growth Powered by Religion, Auto, Education Categories." Watershed Publishing, April 3, 2008. http://www.marketingcharts.com/print/catalog-growth-powered-by-religion-auto-education-categories-4073/ (accessed July 10, 2008).

Marketing Management Analytics (MMA). *Measuring Magazine Effectiveness: Quantifying Advertising and Magazine Impact on Sales.* New York: Magazine Publishers of America (MPA), 2001.

Marshall Marketing & Communications. "Vertis Customer Focus, Retail 2006: Media & Ad Inserts." Vertis Communications, 2006. Summarized at http://printinthemix.rit.edu/summaries/show/24 (accessed March 10, 2009).

McGuire, William J. "An Information Processing Approach to Advertising Effectiveness." In *The Behavioral and Management Sciences in Marketing*, edited by Harry J. Davis and Alvin J. Silk. New York: Ronald Press, 1978: 156-180.

McKibben, Sarah, and Julie Shaffer. *Web-to-Print Primer.* Sewickley, PA: PIA/GATF Press, 2007.

Morgan, Robert M., and Shelby D. Hunt. "The Commitment-Trust Theory of Relationship Marketing." *Journal of Marketing* 58, no. 3 (July 1994): 20-38.

Nash, Edward L., ed. *The Direct Marketing Handbook*. 2nd ed. New York: McGraw-Hill, 1992.

Neisser, Ulric. *Cognitive Psychology*. New York: Appelton-Century Crofts, 1967.

Newspaper Association of America. "NAA Analysis of New Google Research Finds Newspaper Advertising Drives Online Traffic, Consumer Purchasing." NAA news release, April 15, 2008. http://www.naa.org/PressCenter/SearchPressReleases/2008/NAA-ANALYSIS-OF-NEW-GOOGLE-RESEARCH-FINDS-NEWSPAPER-ADVERTISING.aspx (accessed September 15, 2008).

———. "Newspapers: The Source for Bargains." Newspaper Association of America, March 2007. http://www.naa.org/research/newspaper_bargains2007.pdf (accessed September 16, 2008).

———. "Total Paid Circulation." Newspaper Association of America. http://www.naa.org/TrendsandNumbers/Total-Paid-Circulation.aspx (accessed February 6, 2008).

O'Malley, Lisa, and Caroline Tynan. "Relationship Marketing in Consumer Markets: Rhetoric or Reality?" *European Journal of Marketing* 34, no. 7 (June 2000): 797-815.

Olmstead, Larry. "New Styles for Newspaper Leaders: Do You Have What It Takes to Lead Change?" *Newspaper Marketing* (Newspaper Association of America), September/October 2007. http://www.naa.org/Resources/Publications/Newspaper%20Marketing/Newspaper-Marketing-2007-SeptOct/NewspaperMarketing-2007-SeptOct-New-Styles-for-Newspaper-Leaders.aspx (accessed March 14, 2009).

Payne, Adrian, and Pennie Frow. "A Strategic Framework for Customer Relationship Management." *Journal of Marketing* 69, no. 4 (October 2005): 167–176.

Pellow, Barbara A., and Douglas Larsen. "It's Transformation Time for In-Plants." *In-Plant Graphics* 57, no. 12 (2007): 18–20.

Pellow, Barbara A., Michael J. Pletka, and Heather Banis. *Investing in Digital Color... The Bottom Line* (PICRM-2003-10). Rochester, NY: Rochester Institute of Technology, Printing Industry Center, November 2003.

Pellow, Barbara A., Cary Sherburne, and Eve Pedula. *The TransPromo Revolution: The Time is Now.* Weymouth: InfoTrends, 2007.

Pellow, Barbara A., Patricia Sorce, Franziska Frey, Lauren Olson, Katie Moore, and Svetlana Kirpichenko. *The Advertising Agency's Role in Marketing Communications Demand Creation* (PICRM-2003-05). Rochester, NY: Rochester Institute of Technology: Printing Industry Center, October 2003.

Peppers, Don. "Jaguar Uses Mass Marketing to Learn about Individual Customers." *Inside 1to1*, Peppers & Rogers Group, a division of Carlson Marketing Group. January 17, 2005. http://www.1to1.com/View.aspx?DocID=28658 (accessed October 8, 2005).

PODi. *2007 PODi Best Practices in Digital Print Case Studies.* Rochester, NY: PODi, the Digitial Printing Initiative, 2007.

Pressey, Andrew D., and Brian P. Mathews. "Barriers to Relationship Marketing in Consumer Retailing." *Journal of Services Marketing* 14, no. 3 (May 2000): 272–85.

"PRIMIR: Trends and Future for Financial and Transactional Printing, Section VI. Consumer Research Summary." *Print in the Mix*. Rochester, NY: Printing Industry Center at RIT, 2008. http://printinthemix.cias.rit.edu/summaries/show/64 (accessed September 17, 2008).

Publishing & Media Group, The. *Custom Publishing: Opportunities Abound for B-2-B Publishers*. An American Business Media White Paper, January 2003. http://www.americanbusinessmedia.com/images/abm/pdfs/committees/oppb2bpubs2003.PDF (accessed October 8, 2005).

Quotations Page, The. "Quotation Details: John Wanamaker: Half the money I…" QuotationsPage.com. http://www.quotationspage.com/quote/1992.html (accessed March 10, 2009).

Raab, David M. "The Market for Marketing Automation Systems." *Information Management*, December 1, 2005. http://www.information-management.com/issues/20051201/1042146-1.html (accessed March 10, 2009).

Reichheld, Frederick F. *The Loyalty Effect: The Hidden Force behind Growth, Profits, and Lasting Value*. Boston: Harvard Business School Press, 1996.

Reinartz, Werner J., and V. Kumar. "The Mismanagement of Customer Loyalty." *Harvard Business Review* 80, no. 7 (July 2002): 86–94.

Renear, Allen. "Technology of Desktop Publishing." *Computer Sciences*. The Gale Group, Inc., 2002. http://www.encyclopedia.com/doc/1G2-3401200443.html (accessed March 9, 2009).

Reporter Staff. "Hello, You." *Reporter Magazine* 57, no. 30 (May 2, 2008): i. Rochester, NY: Rochester Institute of Technology.

Reuters. "Scarborough Research Releases Newspaper Penetration Report." Thomson Reuters, February 11, 2008. http://www.reuters.com/article/pressRelease/idUS175399+11-Feb-2008+PRN20080211 (accessed September 15, 2008).

Rigby, Darrell, and Dianne Ledingham. "CRM Done Right." *Harvard Business Review* 82, no. 11 (November 2004): 118-129.

Rigby, Darrell K., Frederick F. Reichheld, and Phil Schefter. "Avoid the Four Perils of CRM." *Harvard Business Review* 80, no. 2 (February 2002): 101-109.

Rodgers, Shelly, and Qimei Chen. "Post-Adoption Attitudes to Advertising on the Internet." *Journal of Advertising Research* 42, no. 5 (September/October 2002): 95-104.

Rust, Roland T., Valarie A. Zeithaml, and Katherine N. Lemon. *Driving Customer Equity: How Customer Lifetime Value is Reshaping Corporate Strategy.* New York: The Free Press, 2000.

Schultz, Don E. "Determine Outcomes First to Measure Efforts." *Marketing News* 37, no. 18 (September 1, 2003): 7.

Schwartz, Nelson D. "Bigger and Bigger." *Fortune* 150, no. 11 (November 29, 2004): 147-151.

Sheth, Jagdish, and Atul Parvatiyar. "The Evolution of Relationship Marketing." *International Business Review* 4, no. 4 (December 1995): 397-418.

Six Sigma Web site. "The History of Six Sigma." iSixSigma. http://www.isixsigma.com/library/content/c020815a.asp (accessed October 11, 2005).

Sorce, Patricia. *Sourcing of Corporate Print: Three Case Studies* (PICRM-2005-02). Rochester, NY: Rochester Institute of Technology, Printing Industry Center, January 2006.

———. *Relationship Marketing Strategy* (PICRM-2002-04). Rochester, NY: Rochester Institute of Technology, Printing Industry Center, September 2002.

———, and Kimberly Edwards. "Defining Business-to-Consumer Relationships: The Consumer's Perspective." *Journal of Database Marketing & Customer Strategy Management* 11, no. 3 (April 2004): 255-67.

———, and Barbara A. Pellow. "Amount and Complexity of Personalization in Marketing Campaigns." Presented at the 16th annual Direct Marketing Educational Foundation (DMEF) Educator's Conference, New Orleans, LA, October 2004.

———, and Barbara A. Pellow. *Demand for Customized Communications by Advertising Agencies and Marketing Executives* (PICRM-2003-08). Rochester, NY: Rochester Institute of Technology, Printing Industry Center, October 2003.

———, and Michael Pletka. *Digital Printing Success Models: Validation Study, 2004* (PICRM-2004-06). Rochester, NY: Rochester Institute of Technology, Printing Industry Center, December 2004.

Speier, Cheri, and Viswanath Venkatesh. "The Hidden Minefields in the Adoption of Sales Force Automation Technologies." *Journal of Marketing* 66, no. 3 (October 2002): 98–111.

"Spending in Midst of 3-Year Drop, First Since Depression." *Advertising Age* 79, no. 47 (December 29, 2008): 8.

Stafford, Marla Royne, Eric M. Lippold, and C. Todd Sherron. "The Contribution of Direct Mail Advertising to Average Weekly Unit Sales." *Journal of Advertising Research* 43, no. 2 (2003): 173–79.

Steinberg, Brian, and Megan McIlroy. "4A's Media: How Do We Measure Up." *Advertising Age* 79, no. 10 (March 2008): 1.

Strategic Planning Institute, The. "SPI Home." PIMS (Profit Impact of Market Strategy) Program. http://www.pimsonline.com

Sundar, S. Shyam, Sunetra Narayan, Rafael Obregon, and Charu Uppal. "Does Web Advertising Work? Memory for Print vs. Online Media." *Journalism and Mass Communication Quarterly* 75, no. 4 (Winter 1998): 822–35.

Synovate. "Mail Monitor, Synovate's Credit Card Direct Mail Tracking Service, Celebrates 20th Anniversary." *Mail Monitor*, January 2008. http://mailmonitor.synovate.com/news.asp (accessed March 10, 2009).

———. "US Households Will Receive One Billion Fewer Credit Card Offers in 2008; Synovate Mail Monitor Shows Significant Decline in Offers to High Risk Consumers." *Mail Monitor*, January 2009. http://mailmonitor.synovate.com/news.asp (accessed March 10, 2009).

Syracuse, Amy. "You've Got Greenmail." Waste News, October 27, 2008. *http://wastenews.texterity.com/wastenews/20081027/?pg=11* (accessed March 10, 2009)

Szetela, David. "Irrelevant Ads Breed Turned-Off Consumers." MediaPost's *Search Insider*, March 20, 2008. http://blogs.mediapost.com/search_insider/?p=747 (accessed March 10, 2009).

Tolliver-Nigro, Heidi. "Side-by-Side Test Proves 1:1." *Printing News*, August 11, 2008.

U.S. Environmental Protection Agency (EPA). "General Overview of What's in America's Trash" In *Consumer Handbook for Reducing Solid Wastes*. EPA, March 2002.

U.S. Postal Service. "Lowering the Landfill Levels: Ad Mail's Environmental Impact is Small, Says New Study." Link Online Archive, June 15, 2004. http://liteblue.usps.gov/news/link/2004june15_1.htm (accessed October 7, 2005).

———. "The Household Diary Study: Mail Use & Attitudes in FY 2007." March 2008. http://www.usps.com/householddiary/_pdf/USPS_HDS_FY07_web.pdf (accessed March 9, 2009).

Vakratsas, Demetrios, and Zhenfeng Ma. "A Look at Long Run Effectiveness of Multimedia Advertising and its Implications for Budget Allocation Decisions." *Journal of Advertising Research* 45, no. 2 (June 2005): 241–54.

Vesanen, Jari, and Mika Raulas. "Building Bridges for Personalization: A Process Model for Marketing." *Journal of Interactive Marketing* 20, no. 1 (2006): 5–20.

Veronis Suhler Stevenson, LLC. *VSS Communications Industry Forecast 2008–2012*. New York: Veronis Suhler Stevenson, 2008.

Vruno, Mark. "Print Mail's Digital Links." *Graphic Arts Monthly* 80, no. 5 (May 2008): 38. http://www.graphicartsonline.com/article/CA6560212.html (accessed March 13, 2009).

Wann, David. "Affluenza: Curing the Bug. Five Ways to Fight Junk Marketing." *Denver Post*, September 3, 2002.

Weilbacher, William M. "Point of View: Does Advertising Cause a 'Hierarchy of Effects?'" Journal of Advertising Research 41, no. 6 (November/December 2001): 19-26.

Winer, Russell. "What Marketing Metrics Are Used by MSI Members?" In *Marketing Metrics*, Report No. 00-119, edited by Marion Debruyne and Katrina Hubbard (summarizes the proceedings of the Marketing Science Institute conference, Toronto, Canada, October 5-6, 2000). http://www.msi.org/publications/publication.cfm?pub=21 (accessed March 10, 2009).

Wyner, Gordon A. "The ROI Toolkit." *Marketing Research* 16, no. 3 (Fall 2004): 6-7.

Zahay, Debra L., and Abbie Griffin. "Information Antecedents of Personalization and Customization in Business-to-Business Service Markets." *Journal of Database Marketing & Customer Strategy Management* 10, no. 3 (April 2003): 255-71.

ABOUT THE PRINTING INDUSTRY CENTER AT RIT

The Printing Industry Center at RIT is dedicated to the study of major business environment influences on the printing industry, precipitated by new technologies and societal changes. The Center addresses the concerns of the printing industry through educational outreach, research initiatives, and print evaluation services. The Center serves as a neutral platform for the dissemination of knowledge that can be trusted by the industry, as a place for printing companies and associations to share ideas, and as a meeting ground for building the partnerships needed to sustain growth and profitability in a rapidly changing market.

With the support of RIT, the Alfred P. Sloan Foundation, and our Industry Partners, it is our mission to continue to develop and articulate the knowledge necessary for the long-term economic health of the printing industry.

Printing Industry Center
Rochester Institute of Technology
55 Lomb Memorial Drive
Rochester, NY 14623
http://print.rit.edu

The research agenda of the Printing Industry Center at RIT and the publication of research findings are supported by the following organizations:

INDEX

ADP Dealer Management System, 170, 171
Advertising
    affective responses to, 30
    attention to, 180, 181
    behavioral, 6
    catalog, 17, 25
    clutter, 23, 24
    co-op, 64
    cost per response, 23, 185, 186, 187, 187*tab*
    cost per thousand, 185, 186
    direct, 7, 23, 184–192
    effectiveness measurement, 177
    expenditures, 7, 25*tab*
    exposure, 179
    hierarchy of effects in, 179, 180
    interactive, 184–192
    Internet, 22*fig*, 32*tab*, 33, 48–50
    lifetime value and, 23
    magazine, 27, 28, 29, 32*tab*, 33, 39
    media metrics and, 180–184
    media options in, 20*fig*
    medium congruence, 29
    message comprehension, 180
    newspaper, 17, 22*fig*, 31, 32, 64
    plans, 19
    print, 27–34
    as promotional tool in marketing management, 177
    radio, 17, 22*fig*, 32*tab*
    recall and recognition of, 28
    response per thousand, 187, 187*tab*
    response rates, 31*tab*
    return on investment in, 23
    selective perception and, 180
    on social networks, 48
    television, 17, 22*fig*, 25, 27, 28, 32*tab*, 33, 64
    unwanted, 8
Advertising agencies
    advisory role of, 24
    amount of personalization used by, 38, 38*fig*
    of awareness of new technology, 43, 44, 44*tab*
    creative development by, 21
    responsibilities of, 19
    role in media selection, 21–24, 37
    services provided by, 21–24
Amazon, 151
American Business Media (ABM), 32, 34
Application service provider model, 170
Aprimo, 85
Attention, selective, 1
Automated Direct marketing (ADM), 9, 10, 12
Avery, Chuck, 170

Ballard, Dave, 126, 127
Bradish, Nick, 131
Briggs, Rex, 53
Brodie, Roderick, 70
Bronner, Fred, 29
Budington, Jon, 163, 164, 166, 168

Business-to-business firms (B2B)
    customer base, 19
    decision making in, 33*tab*
    direct sales by, 19
    media research and, 32
    promotion products, 24*fig*
    use of media by, 19, 33*tab*

Calder, Bobby, 28
Carew, John, 40
Case studies
    customer incentive programs, 90-98
    custom publishing, 204-205
    financial statement redesign, 139-142
    in-plant printing services, 121-133
    insert testing, 58, 59
    marketing resource management, 90-98
    multi-channel direct marketing programs, 9-12
    newspaper transition, 158-161
    on power of personalization, 9-12
    printers as digital services providers, 163-173
    publishing transition to digital services provision, 154-161
    use of database records, 90-98
    use of transpromotional tactics, 139-147
ChoicePoint, 7
Communication
    consultants, 197-198
    corporate, 119-148
    customized, 42
    dialogue marketing and, 62
    direct, 48
    electronic, 151
    fully-customized, 66
    interactive, 2
    Internet, 7
    interpersonal, 200
    mail merge, 66
    marketing, 50, 184
    mass, 65
    multichannel, 134, 146
    objectives, 19
    one-to-one, 2
    outcomes, 180
    personalized, 2, 3, 37-50, 65-68
    print, 6, 11
    transaction/transpromotional, 26*tab*, 66
    variable data, 11
    versioning, 66
Coviello, Nicole, 70
Cummings, Twyla, 155, 156
Customer relationship management systems
    relational information processes in, 86, 86*tab*, 87
Customer relationship management systems (CRM), 45, 46, 46*tab*, 69-70, 171
    communication strategies and, 53
    development process, 54*fig*
    dialogue marketing and, 62
    implementation of, 86, 87, 96-98
    information acquisition and, 60, 61
    insight into buying behavior and, 53-56
    knowledge of trigger events and, 54
    leveraging information in, 56
    multichannel integration processes, 54*fig*, 85
    performance assessment and, 54*fig*
    strategy framework, 54*fig*, 65-68
    transaction analysis in, 53, 54
    value creation process, 54*fig*
Customer(s)
    acquisition, 13, 56-61, 65-68
    awareness of products, 19
    capturing behavior of, 42
    demographics, 45
    dialogue with, 2

Customers (*continued*)
  equity, 201, 201*tab*, 202, 203
  identification of, 56
  incentive programs for, 64, 69, 90–98
  lifetime value, 23, 56, 62–65, 63, 65
  list rentals/sales and, 56, 57
  loyalty, 54, 62–65, 69, 199, 200
  personalization preferences, 6
  personalization strategies for development of, 53–82
  post-sale interaction and, 54
  preference changes, 180
  relationship development, 62–65
  retention, 9–12, 13, 56, 61–68, 185
  segmentation, 2, 154
  targeting, 2
  willingness to try product, 19
Customization, 1, 2. *See also* Personalization
  defining, 2
  mass, 2

Dahlén, Micael, 29
Danaher, Peter, 70
Data
  analysis, 2
  cleansing, 48, 99, 165
  collection, 165
  degree of specificity of, 99
  management, 164
  merging, 11
  mining, 3, 46, 48, 63, 98, 99
  quality of, 99
  transaction, 45
  updating, 99
  variable, 50, 88*tab*, 99, 108
Databases
  for capturing/organizing customer information, 13
  components of, 99–100
  customer, 54, 86–89
  de-duping, 102
  demographic profiles in, 54
  fundamentals, 13, 99–107
  management, 135
  for personalization strategies, 85–116
  relational, 100–102
  structure, 100*fig*
  technology, 86–89
  and use of conditional logic, 102–107
Data management system, 53
Dealer Office Xpress (DOX), 170–173
Dell Computers, 2
DesignOnDemand, 170, 171, 172
Destino, Tracy, 108
Dewitz, Adam, 68, 71
Digital services, 151–174
  ancillary services, 152, 153, 153*tab*
  application service provider model and, 170
  need for customer research, 154
  new career trajectories in, 168
  in printing, 163–173
  product development strategy, 151
  in publishing, 154–161
  strategies, 152–153
Direct Marketing Association (DMA), 38
Dobscha, Susan, 65
Document management, 68, 134–135
Do-not-track lists, 6, 60
Dreo, Jery, 163
Dun & Bradstreet, 99
Dyson, Esther, 48, 49

Edwards, Kimberly, 69
Email, 6
  cost per lead, 2
  expenditures, 25*tab*
  marketing, 38
  marketing messages by, 10, 136

Index 225

Email (*continued*)
    personalized, 9–12, 136
    promotions, 25
    response rates, 31*tab*
Enterprise marketing management. *See* Marketing resource management (MRM)
Enterprise resource planning system (ERP), 170
Evaluation. *See* Measurement
Experian, 99

Fay, Kevin, 165, 166
Feedback, design of systems for, 13, 177–195
Fenter, James, 139, 141
Fenton, Howie, 67
First Data Corporation, 142–147
Fisher, Jim, 122, 124
Fitts, Debra, 124, 125
Fournier, Susan, 65
Fox, Shari, 167
French, Bob, 127
Frow, Pennie, 53

Gannett Company, 8, 9–12, 158–161
    creation of new products and revenue streams by, 160
    use of new business models, 156–161
Global Printing, 163–169
Gooding, Elizabeth, 142
Google, 4, 5, 32, 34, 151
    AdWords network, 49
Griffin, Abbie, 2

Hotchkiss, Gord, 3
Hunt, Shelby, 69
Hurwitz, Wendy, 12

The Industry Measure, 38, 39, 43, 45

Information
    access, 86, 86*tab*, 87
    capture, 86, 86*tab*, 87
    cost of obtaining/updating, 87
    integration, 86, 86*tab*, 87
    marketing-related, 10, 11
    personal, 7, 167
    reciprocity, 86, 86*tab*, 87
    transactional, 10, 11, 61
    use of, 86, 86*tab*, 87
Information technology (IT)
    absence of in executing personalized strategies, 13
    infrastructure, 13
    at point of purchase, 2
    requirements for successful implementation, 13
InfoTrends, 38, 42, 43, 134
Inserts, response rates, 31*tab*
Internet
    account management, 9, 12
    advertising, 22*fig*, 32*tab*, 33
    banner fatigue on, 48
    billing/payment on, 61
    changes in information use through, 164
    communication, 7
    consumer-customized web sites, 26*tab*
    customized Web pages on, 3
    as delivery medium, 3
    marketing tactics on, 6
    on-demand, 66
    personalized URLs, 26*tab*
    threat to print publishing, 154

Jayachandran, Satish, 86
Johnston, Wesley, 70

Kalyanam, Kirthi, 62
Kamvar, Sep, 4

Kane, Michael, 155, 158, 159, 161
Kaufman, Peter, 86
Kodak Darwin VI Authoring Tool, 112–114
Kowal, Jason, 166, 167

Ledingham, Dianne, 87
Lippold, Eric, 27
Londres, Dave, 40

Ma, Zhenfeng, 27
Macro, Ken, 134, 135
Mailing(s)
    credit card, 7
    data security and, 7
    direct, 6, 7, 8, 10, 17, 24*fig*, 25*tab*, 27, 31*tab*, 32*tab*, 41, 47*tab*, 61, 64, 152, 186, 187, 200
    environmental issues with, 7, 8
    merges, 188
    personalized, 41
    privacy issues, 7
    response rates, 7, 31*tab*
    unopened, 7
    unsolicited, 7, 8, 60
    zip-code targeted, 64
Malthouse, Edward, 28
Marketing
    budgets, 18
    communications, 50
    customer care programs, 64
    customer preferences, 6
    database, 62
    dialogue, 62, 63, 63*tab*
    direct, 9–12, 23, 24, 25, 57, 87, 146, 184–192
    electronic, 128
    email, 3, 6, 38
    goals, 18, 20*fig*, 65–68
    incentive programs in, 64
    interactive, 23, 62, 184–192
    management, 2, 24–26
    media selection and, 7, 24–26
    objectives, 18–20
    one-to-one, 50, 62
    opt-in, 26*tab*
    personalized, 13, 17–34
    plan, 18–20
    relationship, 62–65, 69–70
    targeted, 170
    targeted database, 26*tab*
    transaction-driven strategies, 70
Marketing/brand management, 68
Marketing communications
    customized, 2
    personalized, 17–34
Marketing executives
    amount of personalization used by, 38, 38*fig*, 41, 41*tab*
    of awareness of new technology, 43, 44, 44*tab*
    media selection decisions by, 37
Marketing resource management (MRM), 85–116
    content distribution in, 85
    planning in, 85
    project management in, 85
    system architecture in, 88*tab*
Marketing strategy, 2
    cross-media, 21
    multichannel, 21, 25
Mathews, Brian, 69
McKibben, Sarah, 68
Measurement
    accountability and, 179
    advertising effectiveness, 13, 179–184
    at boardroom level, 178
    design of feedback systems for, 13, 177–195

Index

Measurement (*continued*)
   metrics for, 178
   with point-of-sale information tracking, 179
   predictive validity, 178
Media
   advertising expenditures, 22*fig*
   age differences in preferences for, 136
   "below the line," 23, 24*fig*
   budgets, 47, 47*tab*
   choices, 17–34
   contextual clues in, 29
   costs, 22
   decision making, 22, 23*fig*
   digital, 33*tab*
   electronic, 6, 37, 161
   expenditures, 25*tab*
   integrated, 17–34
   metrics, 180–184, 185
   personalization and, 13
   planning, 13, 17–20, 21, 185
   platforms, 160
   print, 6
   reach, 181
   selection, 17, 18, 22, 23*fig*, 24–26
   time/space sales, 180, 181
Metrics, marketing performance
   customer retention, 178
   market share, 178
   profits, 178
   sales, 178
Metrics, media
   advertising impact measures, 180–184
   close rates, 185
   cost per response, 185, 186, 187, 187*tab*, 191
   cost per thousand, 185, 186
   customer retention, 185
   email forwarding rates, 185
   lifetime value, 187, 189*tab*, 190*tab*, 191, 192*tab*, 193*tab*, 194*fig*
   net present value, 188, 191
   quality/volume of leads, 185
   response per thousand, 187, 187*tab*, 192
   syndicated services and, 182*tab*
   Web site traffic, 185
Meyer, John, 130
Mick, David, 65
Microsoft, 5
Morgan, Robert, 69

Narayan, Sunetra, 28
Neijens, Peter, 29
Norris, Tim, 9, 11

Obregon, Rafael, 28
O'Malley, Lisa, 70

Pageflex, 110–112
Parvatiyar, Atul, 69
Payne, Adrian, 53
Peck, Adam, 131
Pedula, Eve, 142
Pellow, Barbara, 142
Personalization
   acquisition of new customers and, 56–61, 65–68
   awareness as barrier to adoption of, 43
   as business process, 2
   complexity of, 37, 38–43
   current use as communications tactic, 13, 37–50
   customer demographics and, 45
   customer preferences in, 6, 136, 137
   customer retention and, 61–68
   customized direct mail, 59, 60
   database management, 90–98
   defining, 1–3
   development of strategies for, 13
   early modest use of, 13

Personalization (*continued*)
    effective activities, 199, 199*tab*
    effects of customer reluctance to share information on, 7
    electronic, 48–50
    elements of, 2
    factors impacting levels of, 43–48
    fully-customized, 66
    incentive programs and, 64
    information technology infrastructure requirements, 85–116
    intent/use of data in, 2
    interactive, 64
    levels of complexity, 46*tab*, 47, 47*tab*
    limits to acceptance of, 5
    mail merge use in, 66
    marketing campaigns, 197
    media options, 3–8, 13, 17–34
    need for information infrastructure for, 53
    need for superior customer service, 61–62
    obstacles to use of, 44, 45, 46
    perceived benefits of, 44
    preponderance of simplest types, 42
    privacy issues, 4, 6
    purchase history and, 45
    relationship development and, 62–65
    relevance, 8
    strategies for customer development, 53–82
    techniques, 26*tab*
    transaction/transpromotional tactics in, 66
    use in enhancement of marketing tactics, 41
    use of digital color in, 6, 48
    versioning in, 66
    Web site, 3
Planning
    advertising planning, 18–20, 20*fig*
    integrated media, 13, 17–34
    media, 21, 185, 186
Pletka, Michael, 99
PODi (Print on Demand Initiative), 8, 9–12
Pressey, Andrew, 69
Printers, in-plant, 119–148
    content management in, 134, 135
    corporate, 119–148
    document management and, 134–135
    outside clients for, 124
    print volumes, 123, 124
    revenues, 123, 124, 125*fig*
    transpromotional printing and, 135–138
    as vendor-of-choice, 120
Print(ing)
    ancillary services, 152, 153, 153*tab*
    "clean," 151
    consulting services in, 169
    decline in services, 151
    on demand, 26*tab*, 164, 169
    digital, 40
    digital color, 6, 48
    and digital service provision, 151–174
    diversification in, 152
    efficacy of, 27–34
    growth strategies, 151–152
    impact on market behavior, 29
    integrated with online communication, 163–169
    integrative growth strategy, 151
    as part of integrated media mix, 32
    performance comparisons, 27–34
    personalized, 9–12
    procurement, 68
    product development strategy, 151, 154
    variable data, 26*tab*, 40–41
    web-enabled architectures, 71–82
Printing Industry Center (RIT), 37, 39, 42, 43, 45, 59, 121–133
PrintShop Mail program, 114–115

Product development strategy, 151
Products
    customized, 2
    development of, 18, 155
    distribution, 18, 20*fig*
    medium congruence, 29
    placement for purchase, 18
    price determination, 18
    trial use, 180
Profit Impact of Market Strategy (PIMS), 178
Publishing
    cultural component in, 161
    custom, 24*fig*, 204–205
    decline in print circulation, 156
    digital services and, 153–174
    new business models in, 156–161
    newspaper, 156–161
    outsourced, 24*fig*
    personalized print use by, 39
    traditional model, 156

Raman, Pushkala, 86
Raulas, Mika, 2
Return on investment (ROI), 23
Rigby, Darrell, 87

Sfetko, Jason, 40
Shaffer, Julie, 68
Sharma, Subhash, 86
Shelley, Dan, 131
Sherburne, Cary, 142
Sherman, John, 169
Sherron, Todd, 27
Sheth, Jagdish, 69
Six-Sigma programs, 61–62
Software
    campaign management, 45, 46
    data cleansing, 48
    Kodak Darwin VI Authoring Tool, 112–114

marketing resource management, 85
    Pageflex, 110–112
    PrintShop Mail program, 114–115
    spam-blocking, 60
    variable data printing, 108–117
    XMPie uDirect Classic program, 116–117
Sorce, Patricia, 69
Stafford, Maria, 27
Standard Register Company, 169–173
Stanfast centers, 169–173
Stuart, Greg, 53
Sundar, S. Shyam, 28
Szetela, David, 48

Technology. *See also* Information technology (IT)
    awareness of, 50
    customer database, 46, 86–89
    digital color printing, 151
    disruptive, 13
    electronic verification, 138
    need for awareness of, 43
    for personalization, 85–116
Telemarketing, 6, 30
    response rates, 30, 31*tab*
Thomson Corporation, 154–155
Transpromotional printing, 135–147
    amount of personalization used in, 142
    customer attitudes toward, 136
    implementation, 144–145
    postal rate increases and, 138
    sales process, 143–144
    segmentation in, 142
Tynan, Caroline, 70

Unica, 85
Uppal, Charu, 28
U.S. Postal Service
    direct mail estimates by, 8

U.S. Postal Service (*continued*)
   generational differences in preferece for, 137
   "green" efforts by, 8
   transactional mail in, 135
   use of for marketing messages, 137

Vakratsas, Demetrios, 27
Variable data printing, 130–133, 170–173, 173*fig*
   applications/features, 109–110*tab*
   Kodak Darwin VI Authoring Tool, 112–114
   Pageflex program, 110–112
   PrintShop Mail program, 114–115
   software programs, 108*tab*
   XMPie uDirect Classic program, 116–117
Vertis Communication, 32*tab*
Vesanen, Jari, 2

Wanamaker, John, 177
Web-to-print applications (W2P), 67
   ancillary services, 77, 81
   business transaction complexity, 76, 81
   digital assets input, 73–74, 80*tab*, 81
   distribution, 76–77, 81
   document management, 68
   knowledge/skill requirements, 72–73, 80*tab*, 81
   marketing/brand management, 68
   output intent, 74–75, 81
   print procurement, 68
   product formats, 73, 80*tab*, 81
   proofing, 75–76, 81
   site analysis, 79–82
   software application type, 71–72, 79, 80, 80*tab*
   from Standard Register, 169–173
   system analysis, 71–79
   workflow automation, 68
WellPoint, Inc., 7

Wordekemper, Glen, 143, 144, 145, 147
Workflow automation, 68
XMPie uDirect Classic program, 116–117

Yahoo, 5

Zahay, Debra, 2
Zubak-Skees, Chris, 40
Zweben, Monte, 62

green press
INITIATIVE

RIT Press is committed to preserving ancient forests and natural resources. We elected to print this title on 30% post consumer recycled paper, processed chlorine free. As a result, for this printing, we have saved:

3 Trees (40' tall and 6-8" diameter)
1,582 Gallons of Wastewater
1 million BTU's of Total Energy
96 Pounds of Solid Waste
328 Pounds of Greenhouse Gases

RIT Press made this paper choice because our printer, Thomson-Shore, Inc., is a member of Green Press Initiative, a nonprofit program dedicated to supporting authors, publishers, and suppliers in their efforts to reduce their use of fiber obtained from endangered forests.

For more information, visit www.greenpressinitiative.org

Environmental impact estimates were made using the Environmental Defense Paper Calculator. For more information visit: www.papercalculator.org.